IMAGES
of America

Redondo Beach Pier

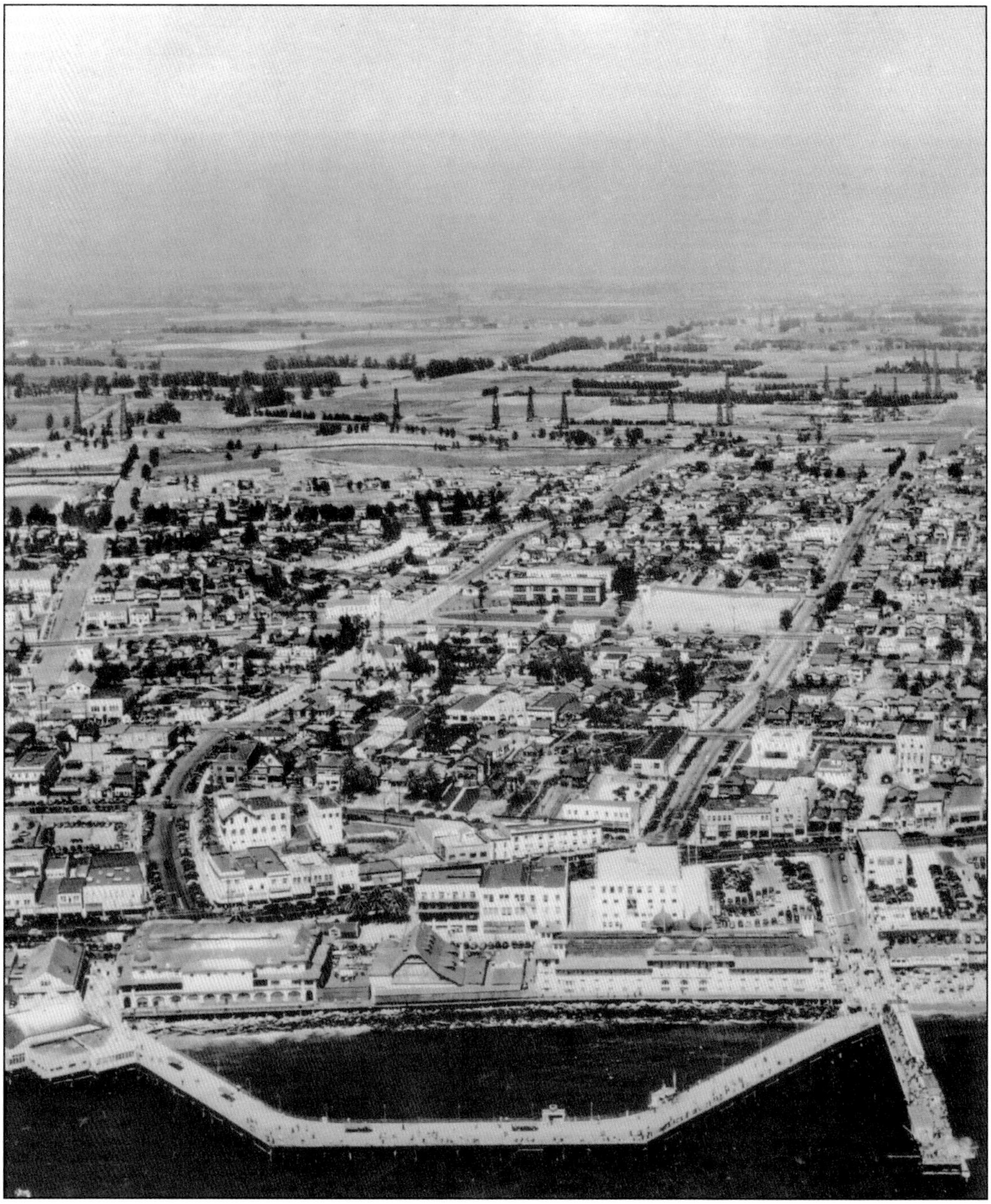

This is Redondo Beach's horseshoe-shaped pier and Monstad Pier in about 1930. At the north entrance of Monstad Pier is the Redondo Beach Pavilion, and at the south entrance is the Redondo Beach Bath House, also called "the Plunge." In the distance can be seen the small city of Redondo Beach as well as the oil fields and rural land that lay inland in the 1930s. (Courtesy of the Redondo Beach Historical Museum.)

On the Cover: This photograph of the Endless Pier in Redondo Beach was taken around 1920. Walking along the Endless Pier was a favorite pastime for visitors and residents alike. Along the outer edge of the pier was a fishing ledge so pedestrians were spared from getting tangled in fishing lines. In the distance is the famous Hotel Redondo. Wharf No. 2 can also be seen just south of the pier. (Courtesy of the Redondo Beach Historical Museum.)

IMAGES
of America

REDONDO BEACH PIER

Jennifer Krintz

ARCADIA
PUBLISHING

ISBN 978-0-7385-7502-5

Published by Arcadia Publishing
Charleston, South Carolina

Printed in the United States of America

Library of Congress Control Number: 2010941328

For all general information, please contact Arcadia Publishing:
Telephone 843-853-2070
Fax 843-853-0044
E-mail sales@arcadiapublishing.com
For customer service and orders:
Toll-Free 1-888-313-2665

Visit us on the Internet at www.arcadiapublishing.com

In honor of my late grandfather, Edwin Moore Krintz,
who taught me the importance of history and family.

Contents

Acknowledgments

I would like to thank the Redondo Beach Historical Museum for the use of its photographs and research. I would especially like to thank my father, Donald Krintz, who is the best research assistant ever. Thank you to Arcadia Publishing, in particular to Jerry Roberts and Debbie Seracini. I would also like to thank Linda Aust at the Redondo Beach Historical Society for her photographs. Thank you to the Historical Commission of Redondo Beach for its blessing of this book, and a very special thanks to Tike Karavas for all of his help, interest, and stories about Redondo in the old days. I'd like to thank my friends who have been wondering where I have been. Last but not least, I would like to thank my family: Grandma, Pop, and Mama, for their encouragement and enthusiasm about this book, and my brothers and sisters, Jon, Amy, Jackie, Mitch, and little Joyce. Stephen, I would also like to thank you for your support, patience, encouragement, and love. I would encourage everyone, especially if you live in Redondo Beach, to visit the Redondo Beach Historical Museum and Morrell House at Heritage Court and to support local history. Unless otherwise noted, all images appear courtesy of the Redondo Beach Historical Museum.

INTRODUCTION

The natural beauty of the Southern California coastline has been drawing tourists for generations. The warm climate, sunny skies, and beautiful white beaches still make Southern California one of the top tourist destinations in the world. Today, these beaches boast pristine sandy shores with modest contemporary, albeit dense, architecture and the occasional unassuming pleasure pier denoting the next beach town along the coast. However, the shoreline of the early 20th century presented a very different culture. Before Disneyland, the pleasure piers in Southern California were the ultimate vacation destination of the early 1900s. Known as the "Coney Island of the West," these pleasure piers of Southern California brought the splendor of the Pacific Ocean and its famous beaches together with the excitement of the American amusement park. Roller coasters, music and dance halls, billiards, casinos, curios, and souvenir shops sat atop these massive piers lining the Southern California coast. The architecture of this time period represented the unveiling of Southern California fame before the peak of the motion picture industry and Disneyland. As the precursor to these famous Southern California attractions, study and exploration of this topic is a necessary chapter in California history.

From 1542 to 1821, Spain ruled what is now California. By the time Mexico declared independence from Spain in 1821, the Spanish governors had distributed several land grants, subdividing the land into several ranchos. Mexican governors took over land grants, and the ranchos continued to subdivide the land. Redondo Beach was part of rancho land granted to Juan Jose Dominguez in 1784. Early Redondo Beach was primarily agricultural up until the late 1800s. With the arrival of the railroad in Los Angeles in 1887, more and more people traveled across the country to settle in the bucolic, fair-weathered landscape and climate of Southern California. Of course, one of the main attractions to this area was its picturesque setting of a pristine white shoreline.

During the late 19th century, the Industrial Revolution played a significant role in boosting the United States as a major player in foreign trade and commerce. The railroad was integral to moving goods and passengers across the country, establishing towns along the way. Along the coasts, ports were set up to import and export freight and people. In California, major ports were built in San Francisco and San Diego. Situated inland, the city of Los Angeles relied on its coastal cities to transport goods and passengers. At one point in the late 19th century, the burgeoning city of Los Angeles had three major ports of call: San Pedro, Santa Monica, and Redondo Beach.

Redondo Beach had three wharves before the construction of the horseshoe-shaped pier, the fourth- or fifth-generation descendant of which can be seen today. The shipping industry in the late 19th century was so successful in Redondo that the town incorporated into a city in 1892. Because of the frequent passenger traffic from the shipping ports, the Redondo Beach Company constructed the beautiful and lavish Hotel Redondo, located on the bluff with a magnificent view of the ocean. This was one of the first beach resort attractions and a sign of the abundant tourism that thrived in later years. Although Redondo Beach had a relatively small shipping port operation, it flourished as a shipping port town up until 1912, when the Los Angeles Harbor in

San Pedro finally took the last of Redondo Beach's shipping business. As a result, the city of Redondo Beach turned its sights to another asset of the beachfront: the beach resort. Because its wharves were no longer in major competition for transport duties, the pleasure pier became the main seaside commodity. Developers began to plan the cities that would appeal to potential commercial investors, residents, and tourists. A pavilion, bathhouse, casino, and hotel drew tourists to Redondo Beach via the Pacific Electric Red Cars line, of which Redondo Beach was a main tourist stop. Under the direction of Henry Huntington, Redondo Beach's shoreline was transformed in the early 20th century with entertainment venues, roller coasters, lavish architecture, and of course, its pleasure piers jutting out into the ocean.

The predecessor of today's Municipal Pier in Redondo Beach is Redondo Beach's Endless Pier (later called Horseshoe Pier), first constructed in 1915. The first pier was simple in design and lacked the entertainment venues that are on the pier today. The natural appeal of the ocean drew a number of tourists that frequented the pier despite its modest design. In those days, ocean activities such as swimming and surfing were done only by masters. Bathhouses were constructed on the beach as a way to swim in the ocean without actually being in the ocean; they were accurate substitutions complete with saltwater baths. For the truly adventurous, a long, thick rope was attached to the bathhouse, and people could walk into the water while holding onto the rope for safety. The treachery of the ocean was still frightening and mysterious to several. The notion of a pier made solely for the purpose of walking above the water had a tremendous appeal to people, possibly as a metaphor for humanity conquering the sea. The innovative horseshoe-shaped design of the pier was an added attraction that made Redondo Beach a unique beach resort town in Southern California. In 1928, the Monstad Pier was constructed on the south side of Horseshoe Pier. It was a straight pier, mainly used for fishing.

Over the years, the pier has been subject to storms and fires and has been constructed and reconstructed. Nonetheless, the early-20th-century beach resort era showcases Redondo Beach as a major tourist attraction. Today, Municipal Pier is a newer generation of these two original pier designs of the early 20th century. Municipal Pier is still in the shape of a horseshoe, and the Monstad Pier is still there in design, though it has been connected at the outer end to the main pier. Although Redondo Beach is now a quiet beach resort town, the pier stands as a reminder of the bustling heyday that once adorned the shoreline.

One

The Redondo Beach Wharves

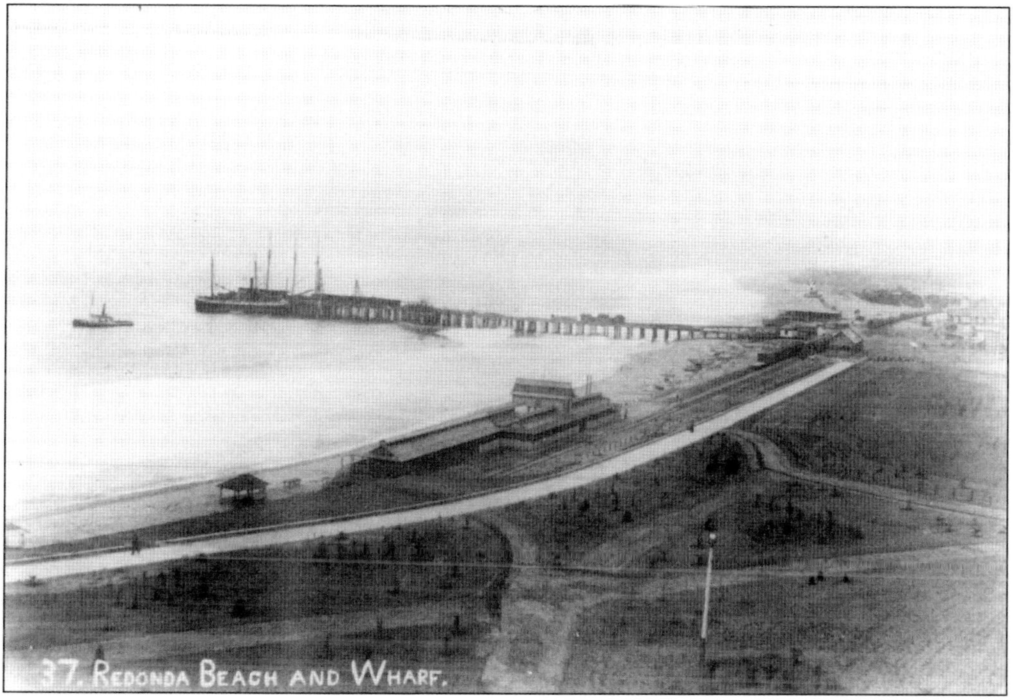

In 1887, the Dominguez Estate Company sold more than 400 acres of beachfront land for $12,000 to the Redondo Beach Company, a group of real estate speculators. With the introduction of the railroad to the town in 1888, as well as the discovery of the deep submarine canyon off of the coastline that allowed large ocean liners to transport freight and passengers via the early wharves, the Redondo Beach Company set out to establish Redondo as a major shipping port for the city of Los Angeles.

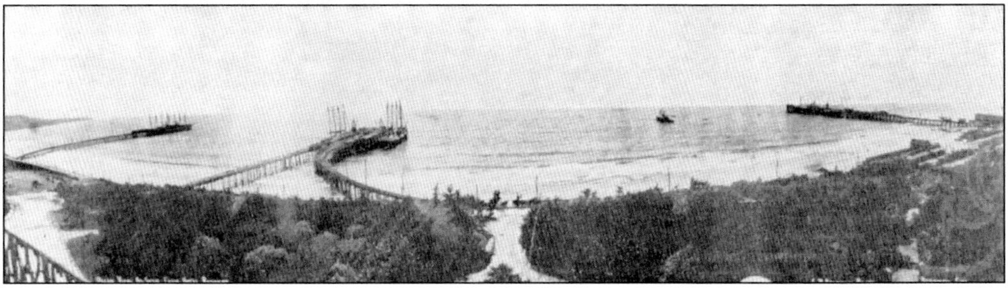

Constructed from 1887 to 1904, three wharves jutted out from the Redondo Beach shoreline at the base of Emerald Street, Sapphire Street, and Ainsworth Court. By the end of the 19th century, industrialization made way for the United States to become a major player in foreign commerce and trade. Shipping ports became integral for transporting freight and passengers. Above are the three wharves of Redondo Beach. From left to right are Wharf No. 3, constructed in 1904 at the end of Sapphire Street; Wharf No. 2, constructed in 1895 at the end of what was once Ainsworth Court in front of Hotel Redondo; and Wharf No. 1, constructed in 1888 at the end of Emerald Street. Below is Wharf No. 1 shortly after construction in 1889.

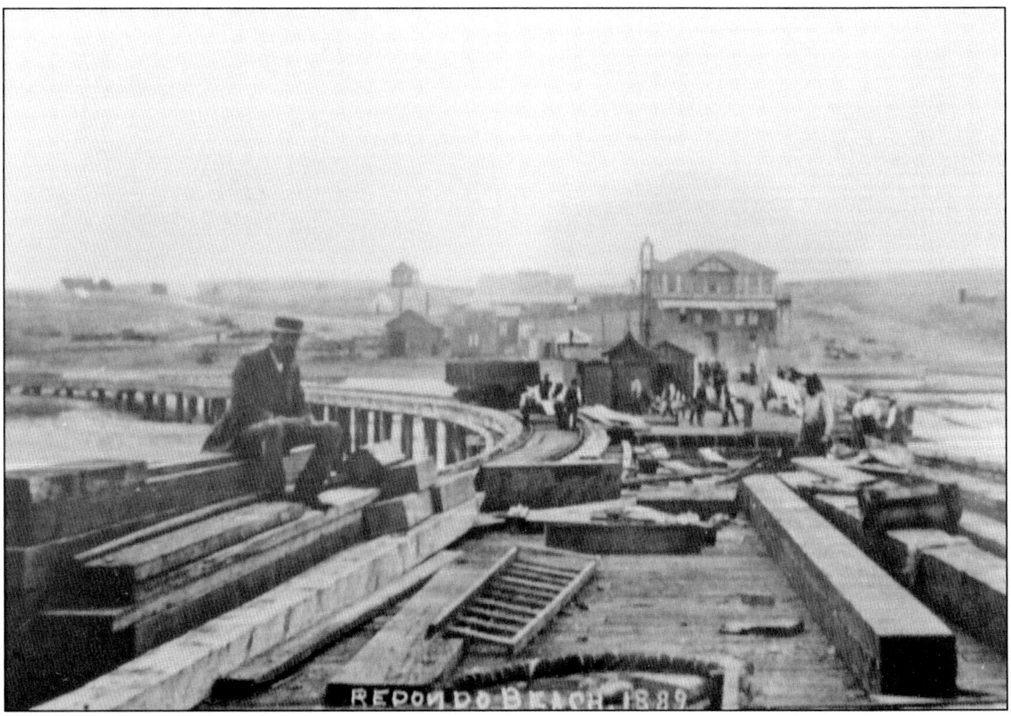

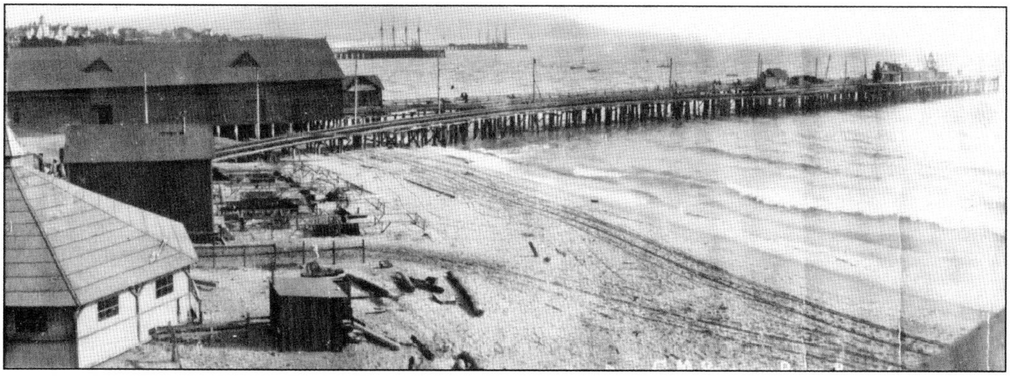

Above is a photograph taken in the early 1900s from the north, facing south. In the background is the Palos Verdes Peninsula. The photograph below faces north toward the three wharves from the south. These three wharves helped make Redondo Beach one of the contenders for the Los Angeles Harbor in the late 1800s. Other possibilities included Santa Monica's Long Wharf to the north and San Pedro to the south. San Pedro eventually won the battle of the ports, but Redondo Beach stayed in the shipping industry until San Pedro took the last of Redondo's business shortly after 1912.

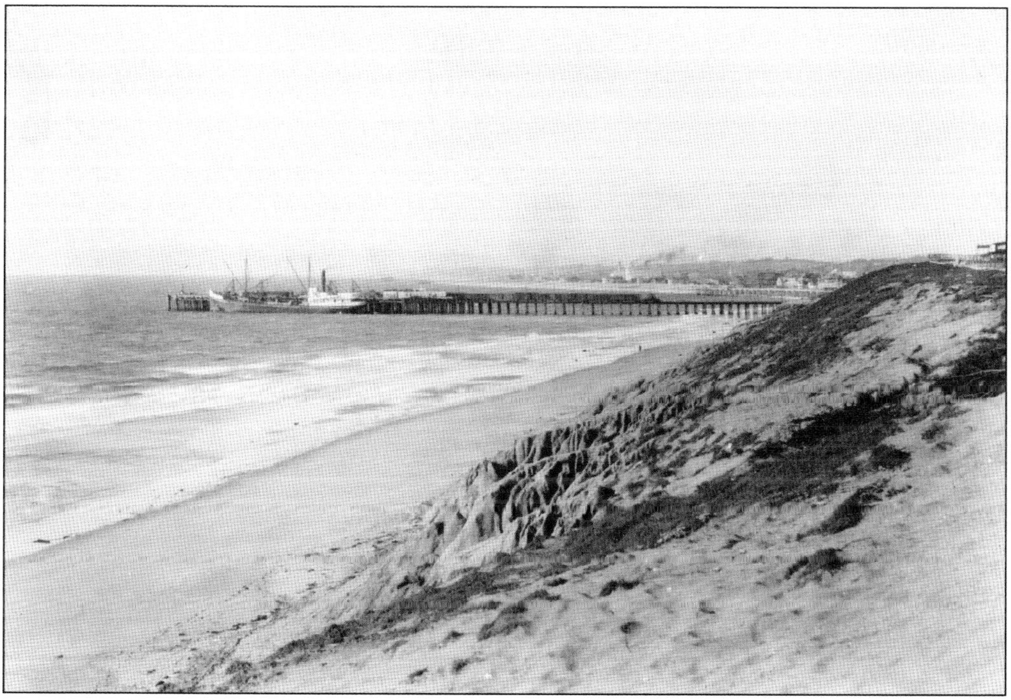

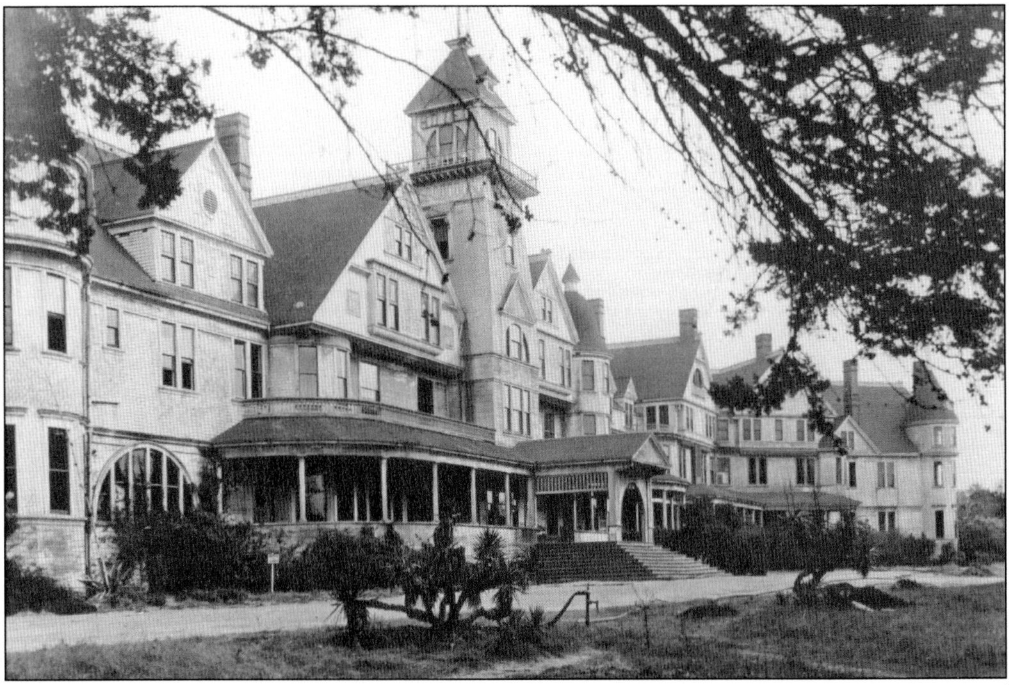

As part of the shipping industry, tourist and passenger traffic passed through the small town of Redondo Beach in the late 19th century. To lure these passengers to stay and enjoy the seaside, the Redondo Beach Company constructed the Hotel Redondo in 1890. It was a beautiful hotel built along the bluffs, overlooking the wharves and the Pacific Ocean. This hotel was one of the most extraordinary of its time. Erected at the same time as the Hotel Del Coronado in San Diego, Hotel Redondo boasted 225 rooms, baths on every floor, a golf course, tennis courts, and a grand ballroom.

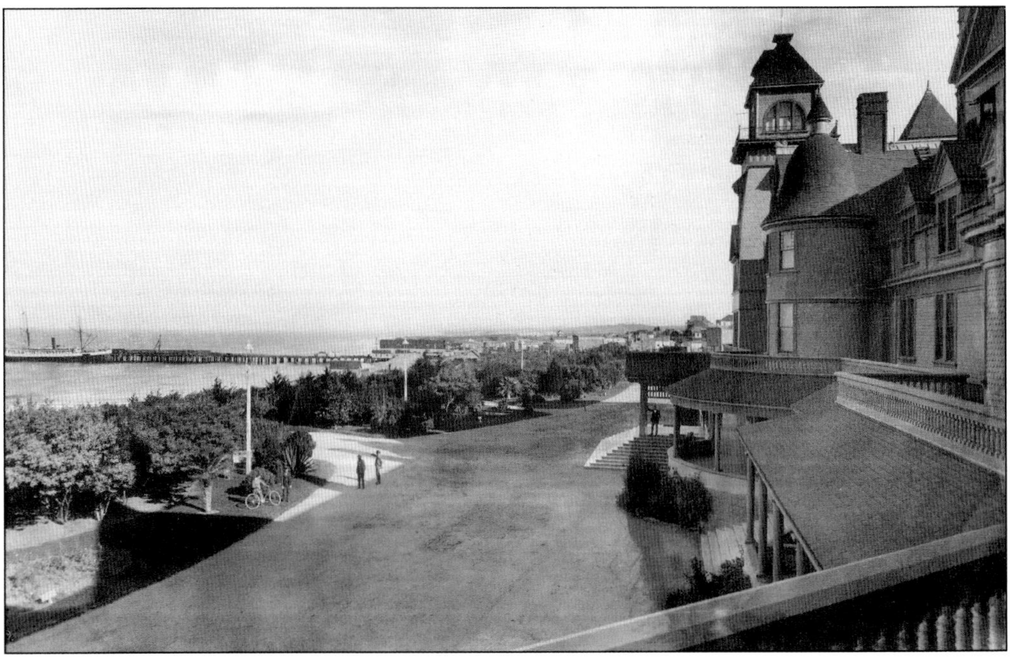

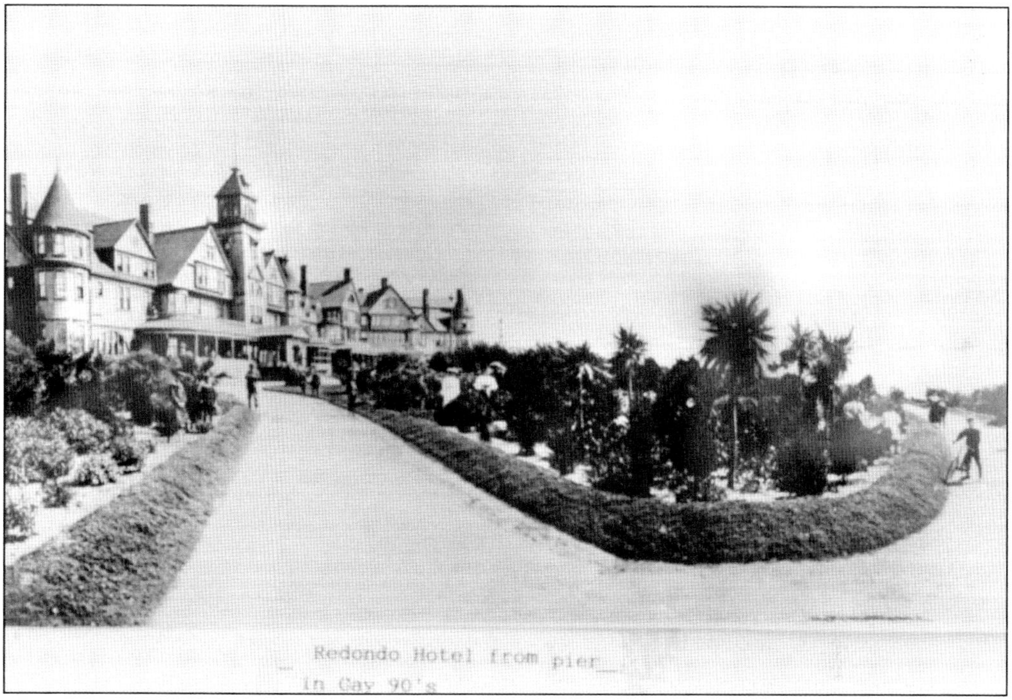

Shown here is Hotel Redondo in the late 1890s. The hotel was constructed in a High Victorian style on a towering bluff overlooking the Pacific Ocean. It was a three-story hotel that had a complex multi-gable roof adorned with several dormer windows and turrets. The photograph above captures the beautiful landscaping in front of the hotel that led to the beach. In the image below, people enjoy a round on the golf course adjacent to the hotel.

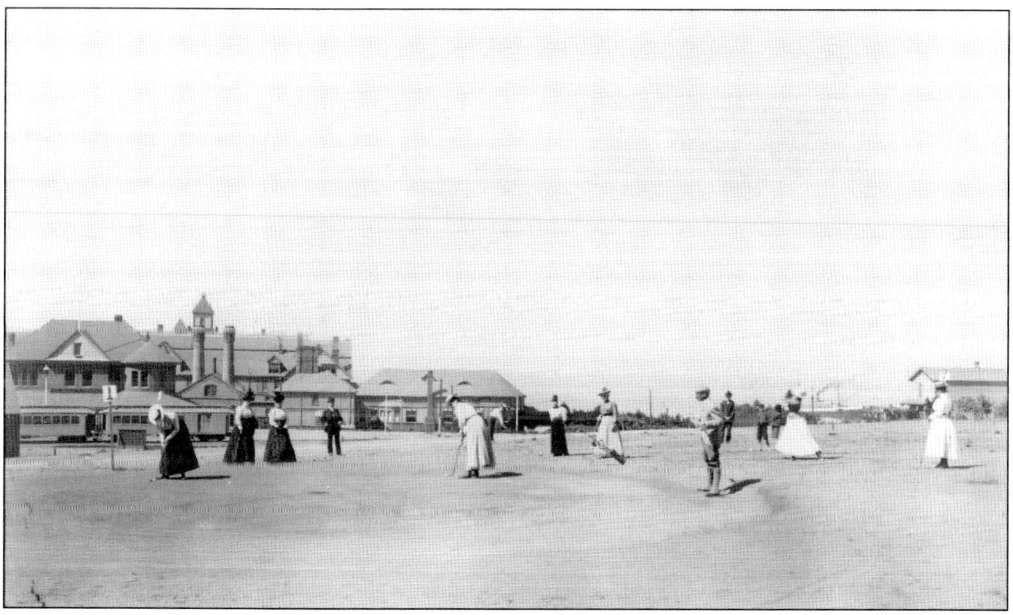

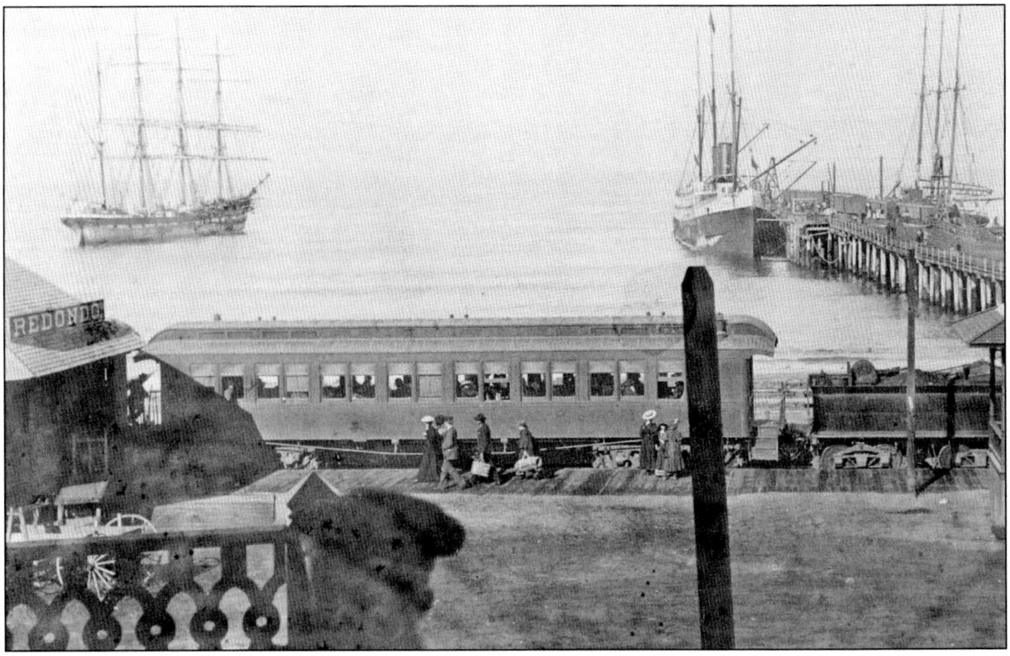

The Redondo Beach Company was headed by two gentlemen, J.C. Ainsworth and R.R. Thompson, who were both Oregon lumber and shipping masters. They renamed the business the Redondo Improvement Company. Two railway companies serviced Redondo Beach: the Santa Fe Railway and the Redondo Railway, which was established by the Redondo Improvement Company. The Redondo Railway focused on tourist and passenger transport, while the Santa Fe Railway concentrated on commercial shipping needs.

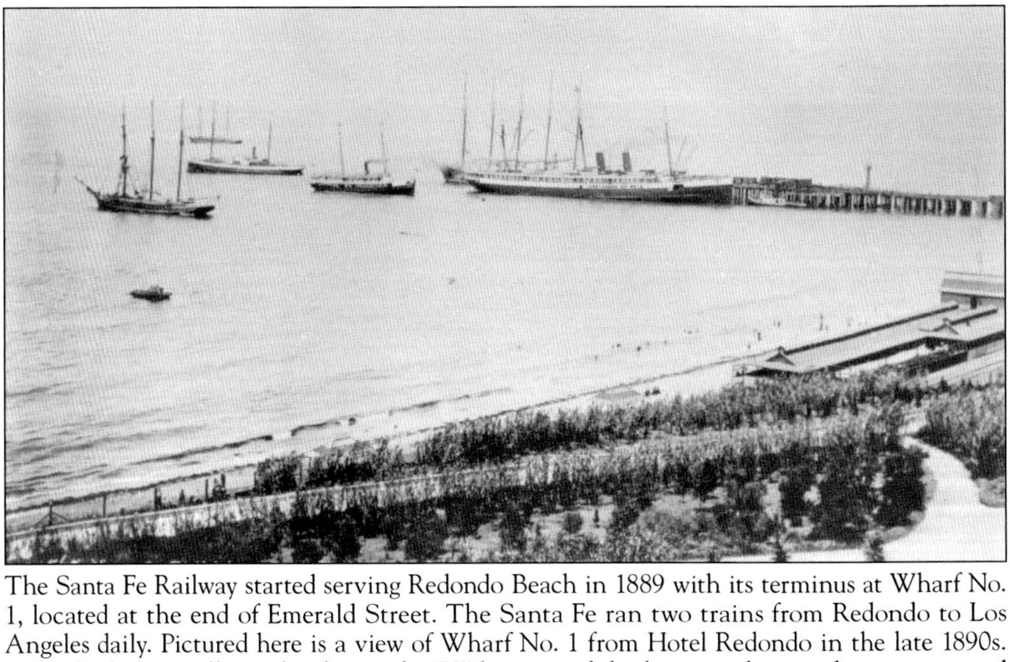

The Santa Fe Railway started serving Redondo Beach in 1889 with its terminus at Wharf No. 1, located at the end of Emerald Street. The Santa Fe ran two trains from Redondo to Los Angeles daily. Pictured here is a view of Wharf No. 1 from Hotel Redondo in the late 1890s. It was built originally in the shape of a "Y," but one of the legs was destroyed in a storm and never reconstructed.

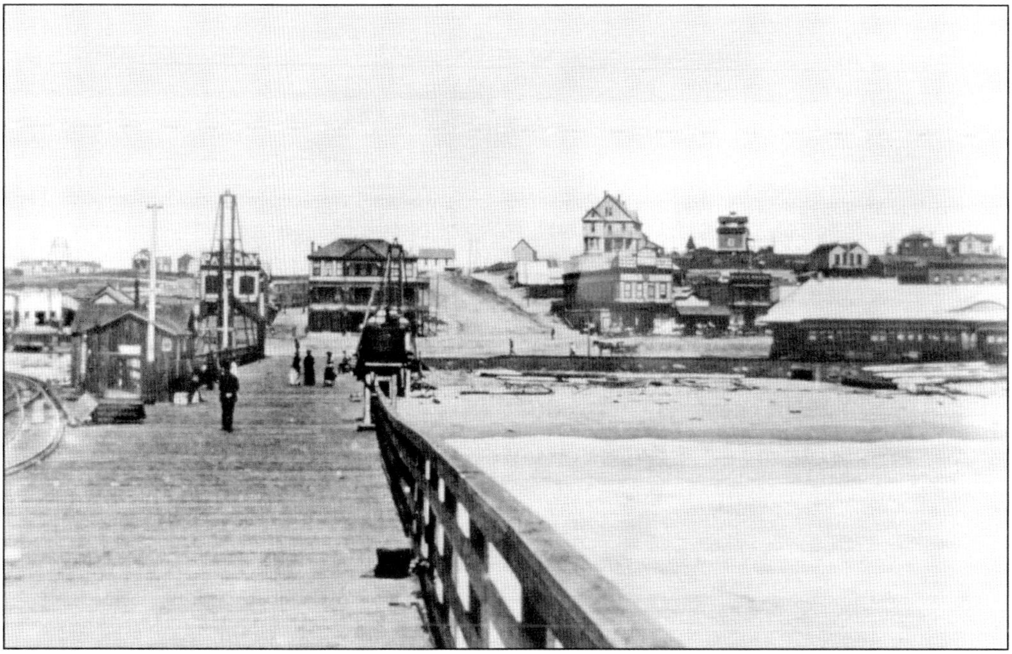

Los Angeles grew tremendously in the late 19th century. The railroad transported an influx of immigrants, Midwesterners, and East Coasters that made the city grow exponentially. This is a photograph of Redondo Beach from Wharf No. 1 taken around 1900. The city of Redondo Beach has a few houses along the hills in the background.

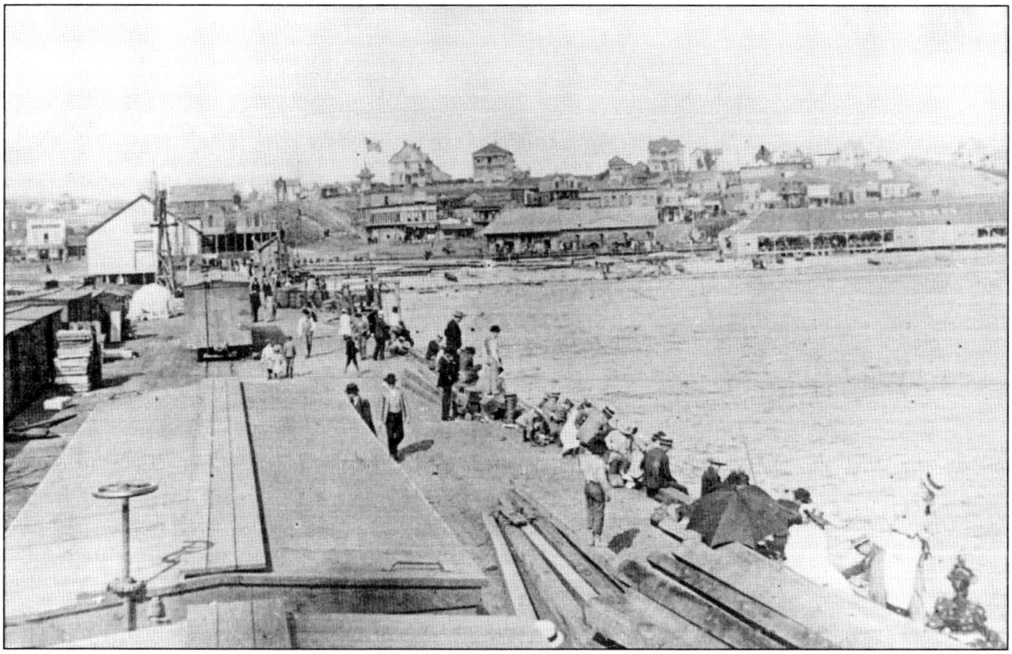

This is the same scene a few years later, taken a few feet back on the wharf. Note the people fishing. Although this wharf was constructed for shipping port capabilities, people still ventured to it for a promenade on the wharf or to sit down to fish. Fishing in Redondo has always been a form of recreation for visitors and residents alike.

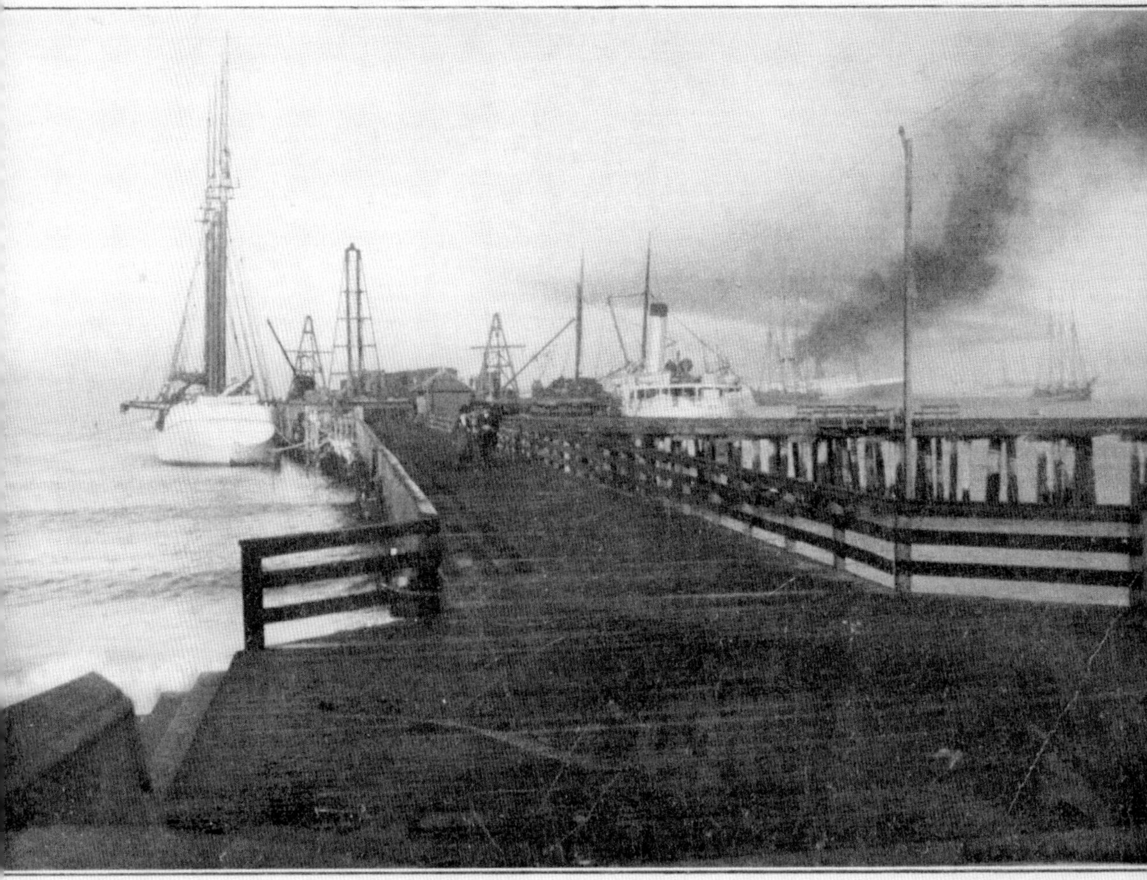

WHARF NO. 2—REDONDO.

Wharf No. 2 was constructed in 1895 during the battle for the Los Angeles Harbor between Santa Monica's Long Wharf, San Pedro, and Redondo Beach. Wharf No. 2 was built at the end of Ainsworth Court, located approximately in front of Hotel Redondo. It was designed in the shape of a "Y." One leg was tracked for trains, and the other leg was for fishermen and pedestrians.

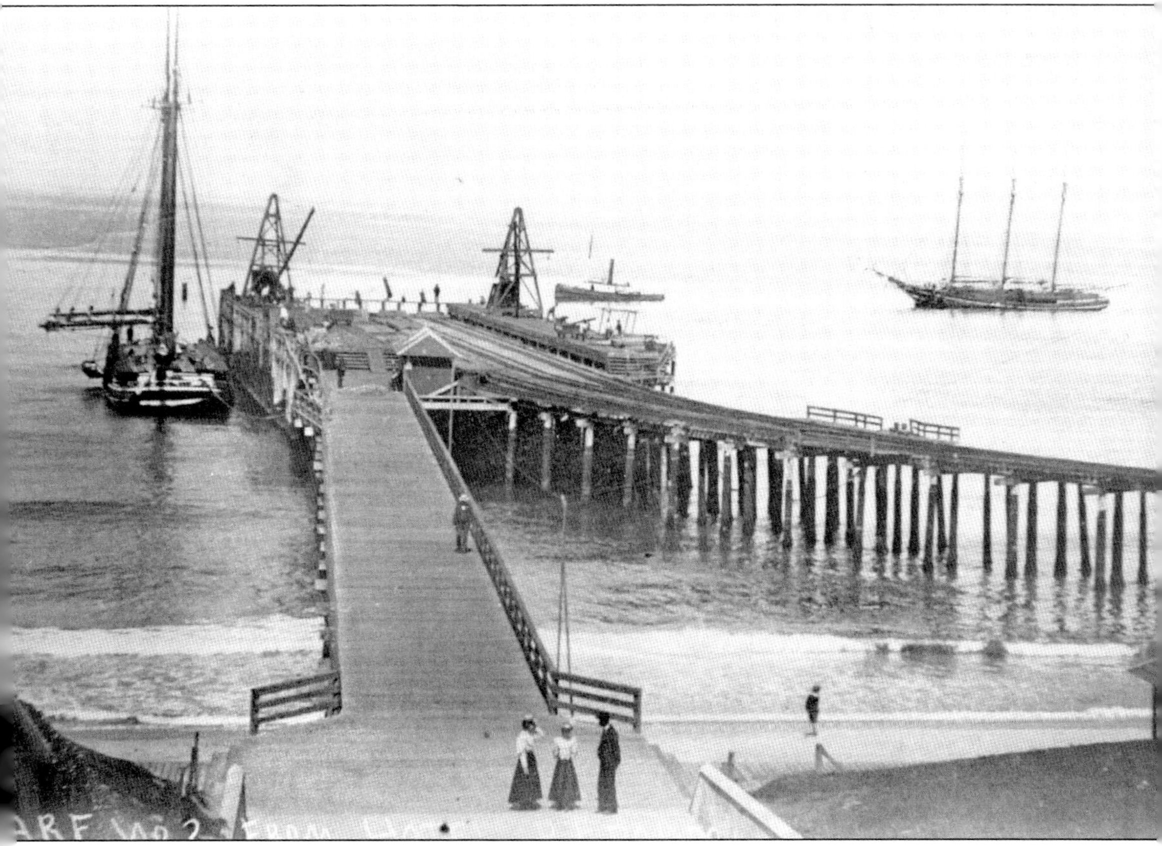

A year after Wharf No. 2 was constructed, the Redondo Railway Company was reorganized and became the Redondo & Los Angeles Railway. Wharf No. 2 was eventually dismantled because it had become unusable from sand silting in 1916.

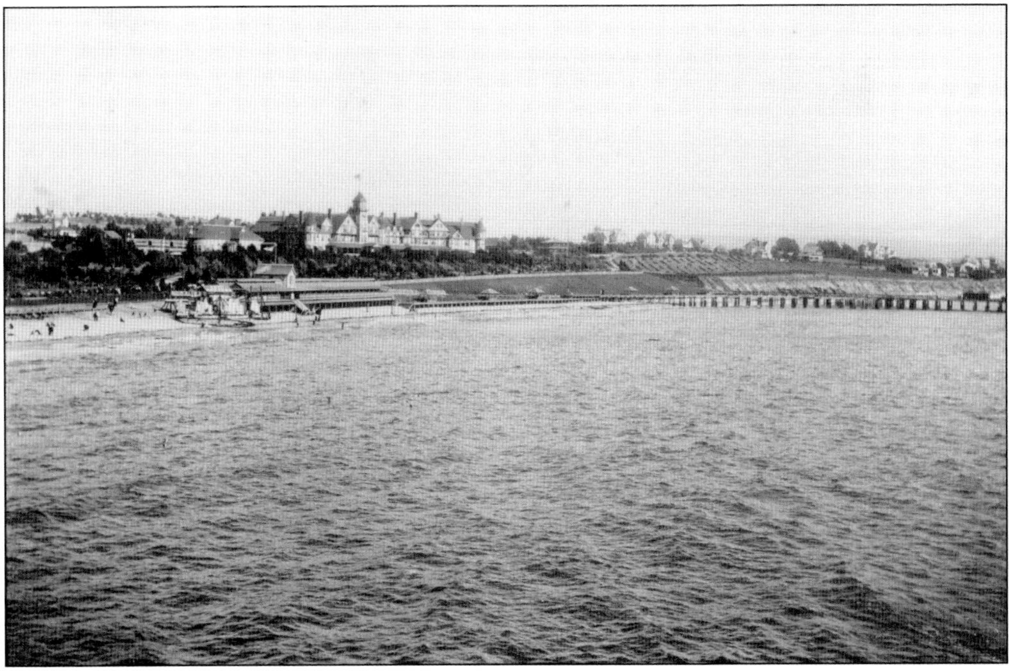

This photograph of the beach and Hotel Redondo was taken from Wharf No. 1 in the late 1800s.

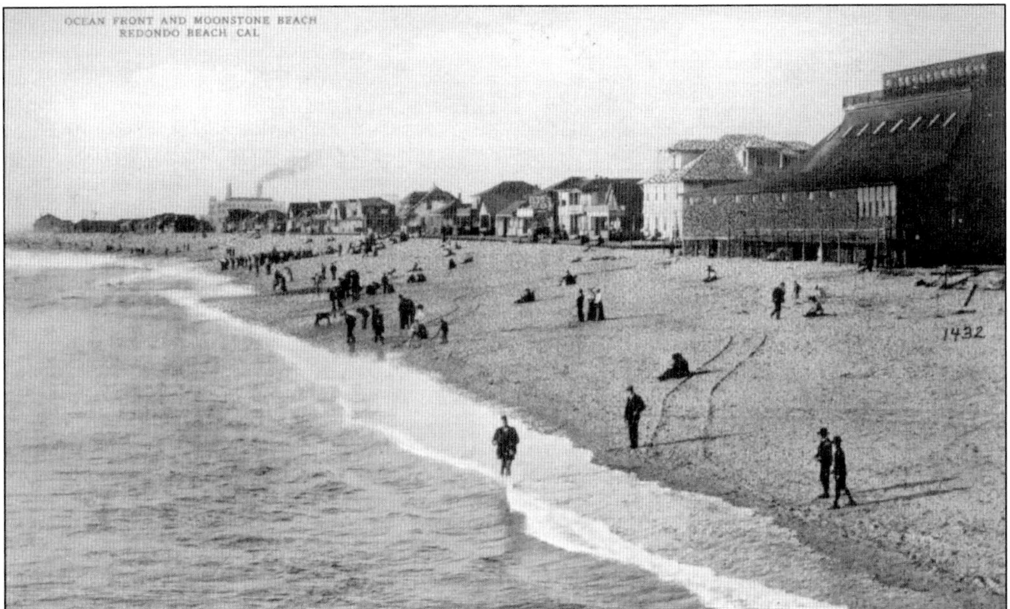

One of the natural tourist attractions in Redondo Beach was Moonstone Beach, located north, next to the Hermosa Beach border. Moonstone Beach was given its name because of the natural gemstones, such as moonstones, that were found along the shore. Moonstone Beach is no more because the beautiful stones were eventually all taken before government intervention could preserve the natural beauty of the site. Today, remnants of this beach can be found in historic postcards and photographs. Above is a postcard of Moonstone Beach in Redondo. (Courtesy of Donald Krintz.)

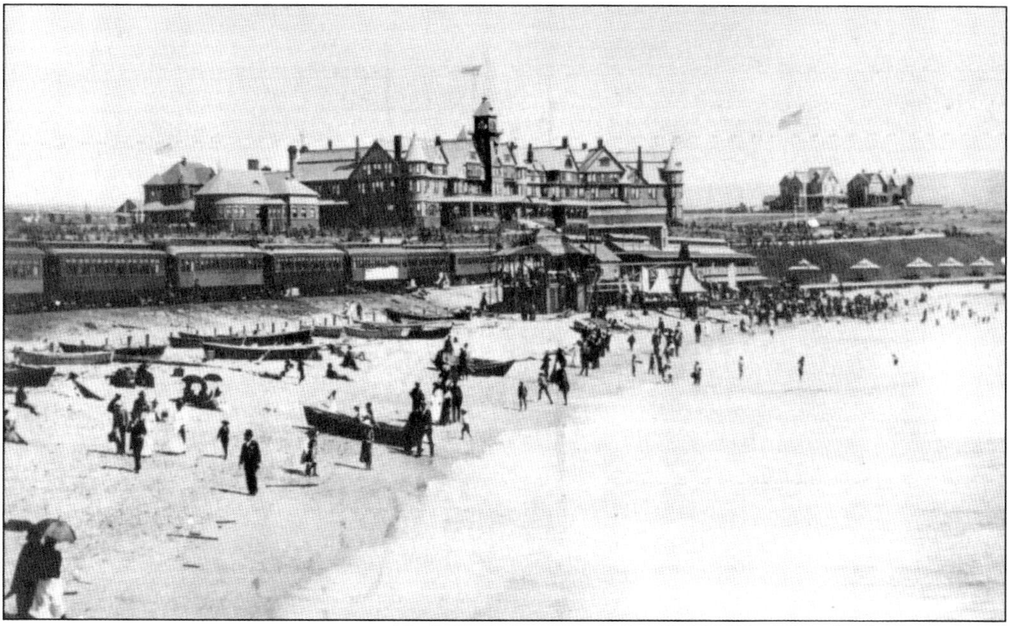

On April 18, 1892, the town of Redondo Beach incorporated as a city in Los Angeles County. Around this time, the Redondo Improvement Company built the first pavilion, a bathhouse, the Hotel Redondo, and other entertainment venues such as a bandstand and casino. While the town's mainstay was still embedded in the shipping business, it did not distract from the natural draw of the seaside. These photographs show the precursory buildings and venues to the exotic fantasyland architecture that was later introduced to the Redondo Beach shoreline. The image above showcases the beautiful Hotel Redondo overlooking the beach, and below is a view of the section of beach just north of the Hotel Redondo.

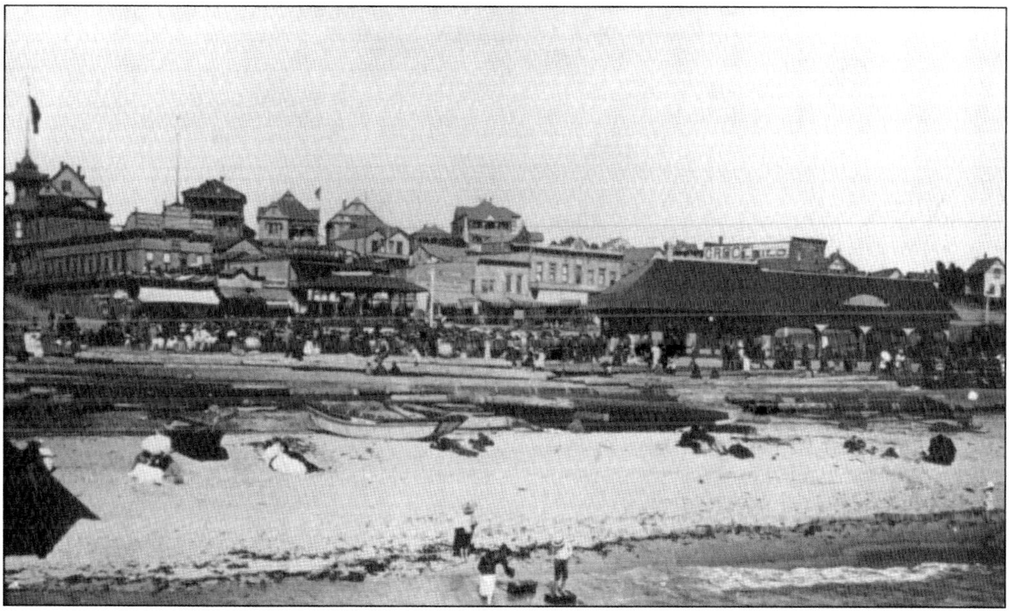

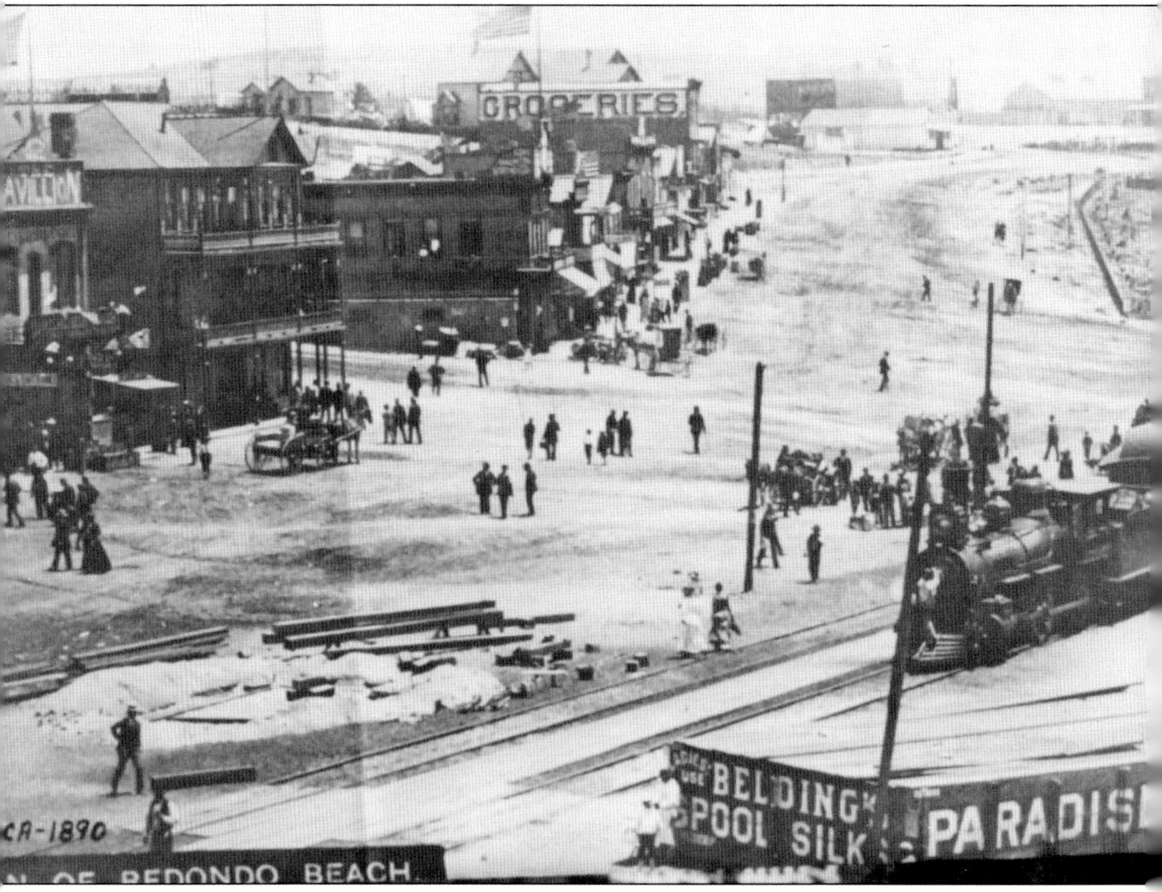

This is the small city of Redondo Beach with an unpaved Pacific Avenue (now primarily condominiums) in the background. In the foreground of the photograph is a steam locomotive that is most likely carrying freight from one of the wharves. Steam engines were used until 1902, when railroad companies switched to electric power.

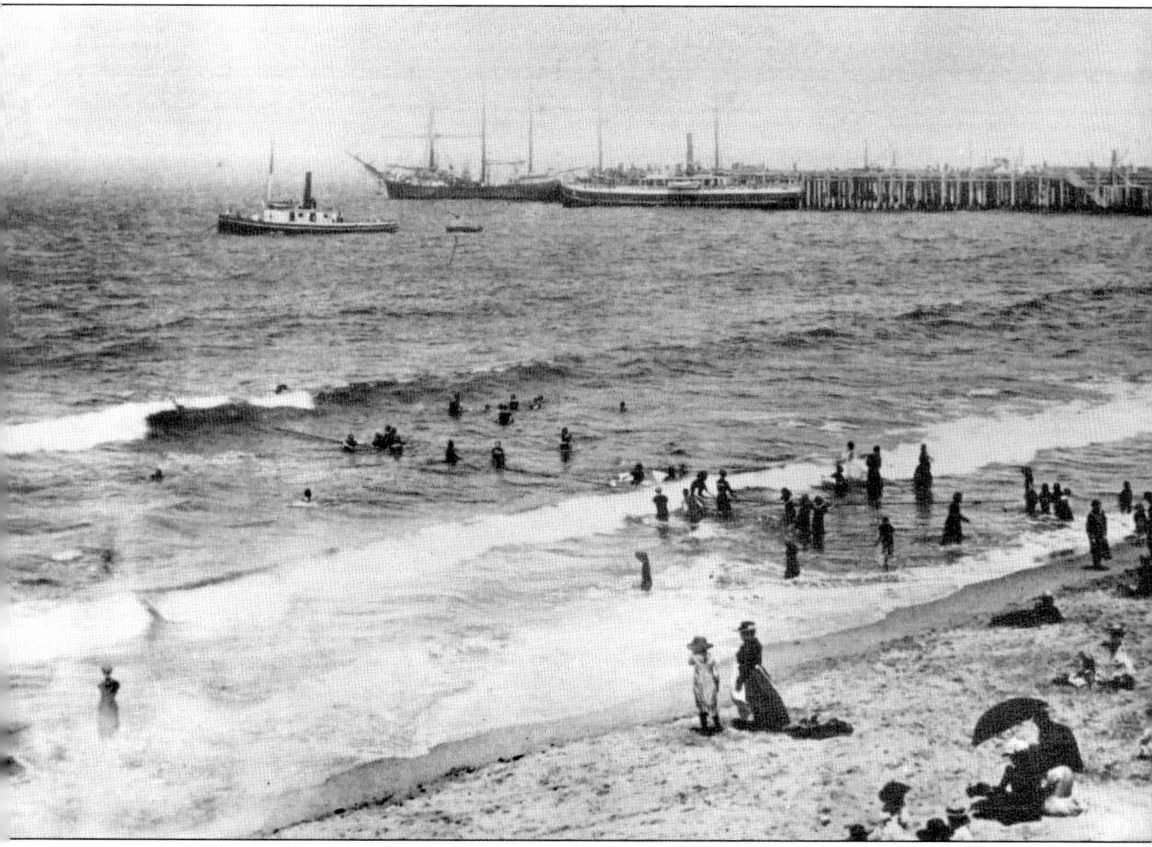

In this photograph are bathers at Redondo Beach in the 1890s. This was likely taken in front of the first bathhouse in Redondo Beach, which was a modest wood-clad structure compared to the later extravagant bathhouse constructed under the direction of Henry Huntington.

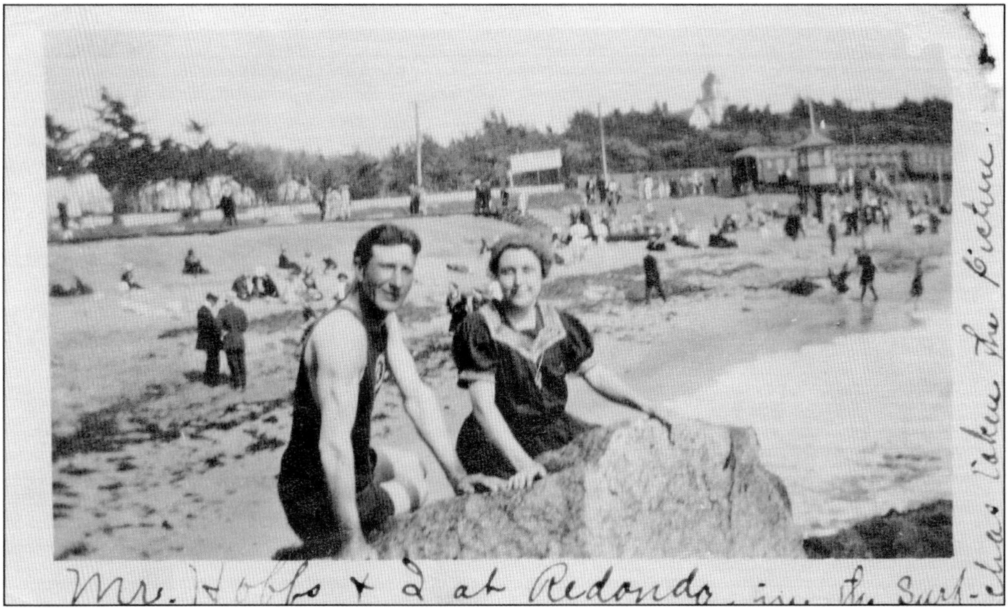

Seen above are two bathers at Redondo Beach in the early 1900s. In the background, the grand Hotel Redondo can be seen. Typically, bathing suits could be rented from the Bath House, which is why many of the bathing suits look so similar to one another at this time.

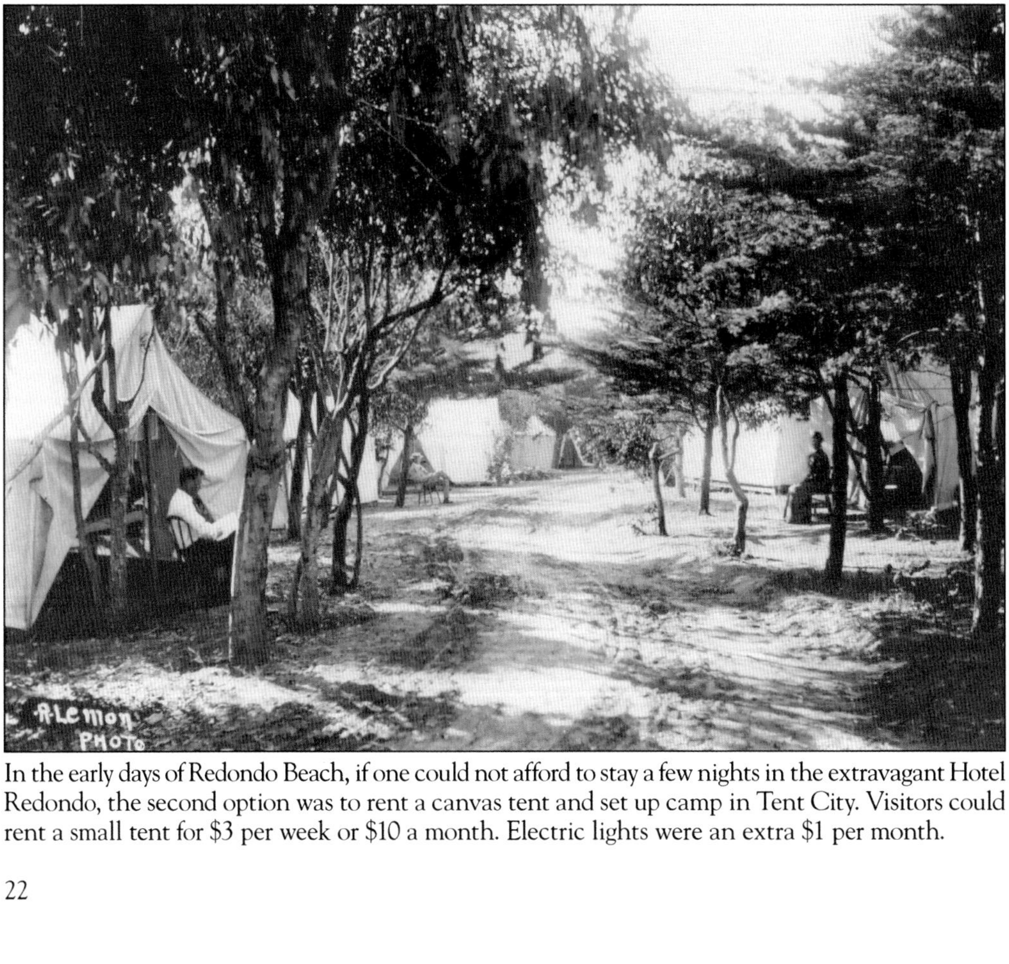

In the early days of Redondo Beach, if one could not afford to stay a few nights in the extravagant Hotel Redondo, the second option was to rent a canvas tent and set up camp in Tent City. Visitors could rent a small tent for $3 per week or $10 a month. Electric lights were an extra $1 per month.

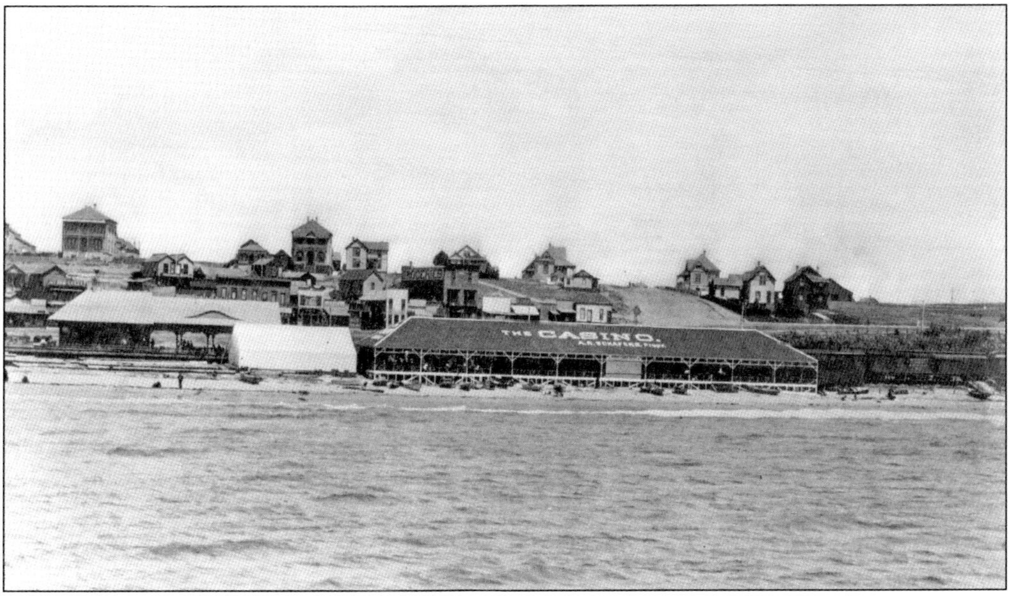

This is a view of the beach from either a boat or the end of Wharf No. 1 taken around 1900. The large building on the beachfront is a casino that was later replaced by the El Paseo entertainment area and other amusements that adorned the shoreline in the early 1900s.

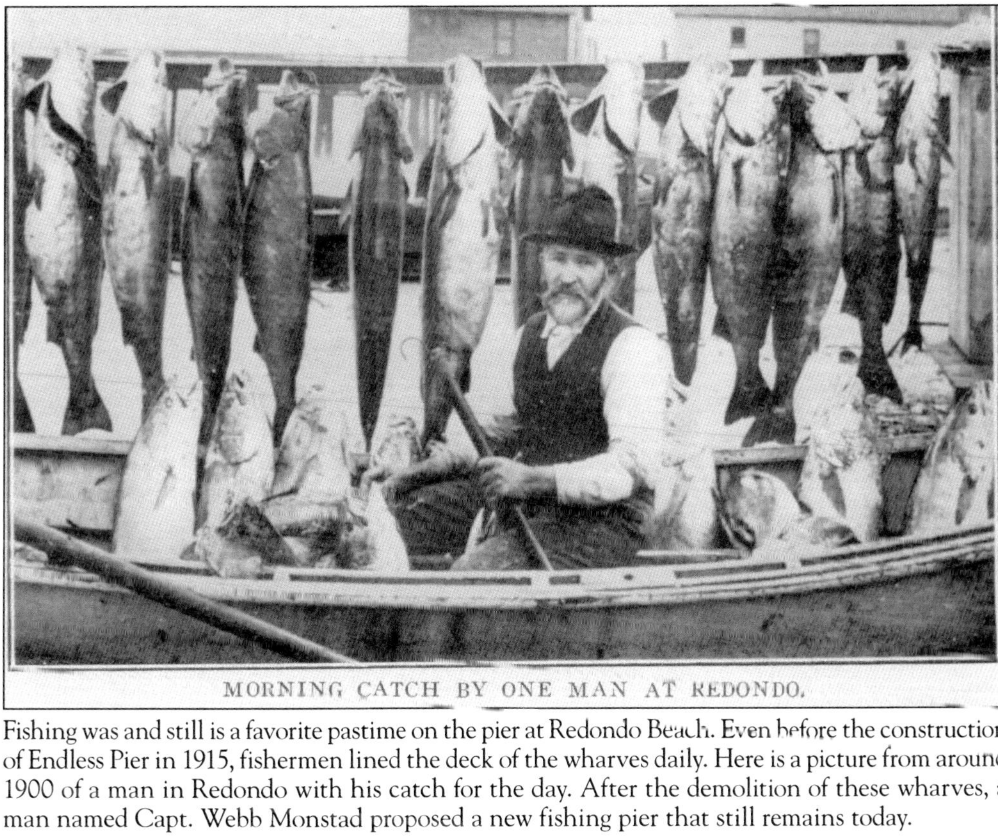

MORNING CATCH BY ONE MAN AT REDONDO.

Fishing was and still is a favorite pastime on the pier at Redondo Beach. Even before the construction of Endless Pier in 1915, fishermen lined the deck of the wharves daily. Here is a picture from around 1900 of a man in Redondo with his catch for the day. After the demolition of these wharves, a man named Capt. Webb Monstad proposed a new fishing pier that still remains today.

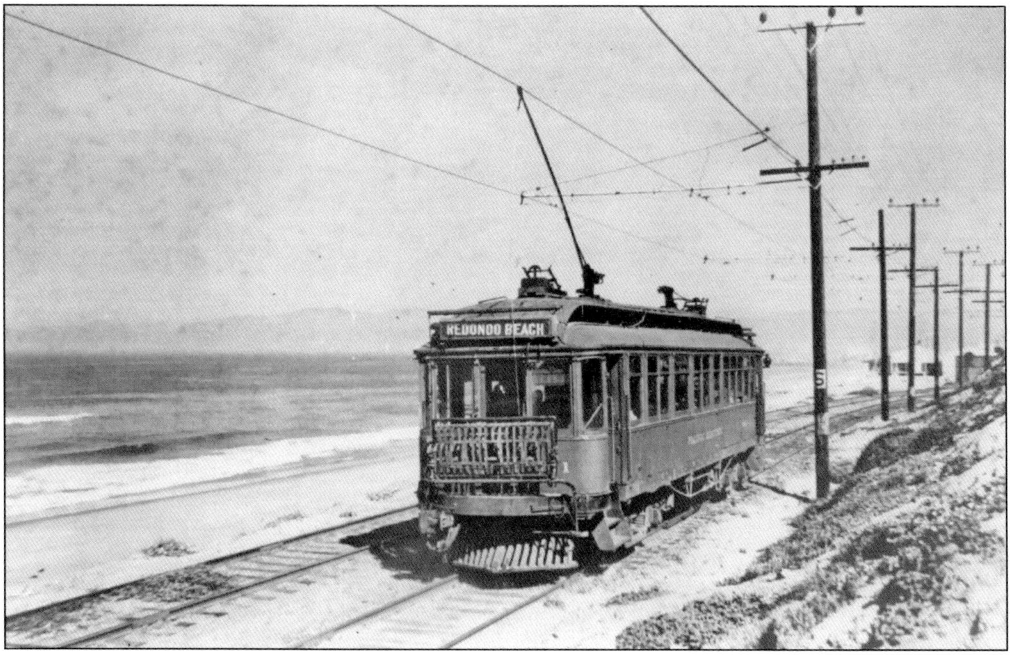

In 1905, Henry Huntington, a real estate and railroad tycoon, bought the Los Angeles & Redondo Railroad track for his new Pacific Electric Red Car. He also bought the Redondo Improvement Company and set out to market Redondo Beach to potential residents and business owners. As history illustrates, Huntington's focus was on promoting the beach resort aspect of Redondo. Pictured is one of Huntington's Pacific Electric Red Cars.

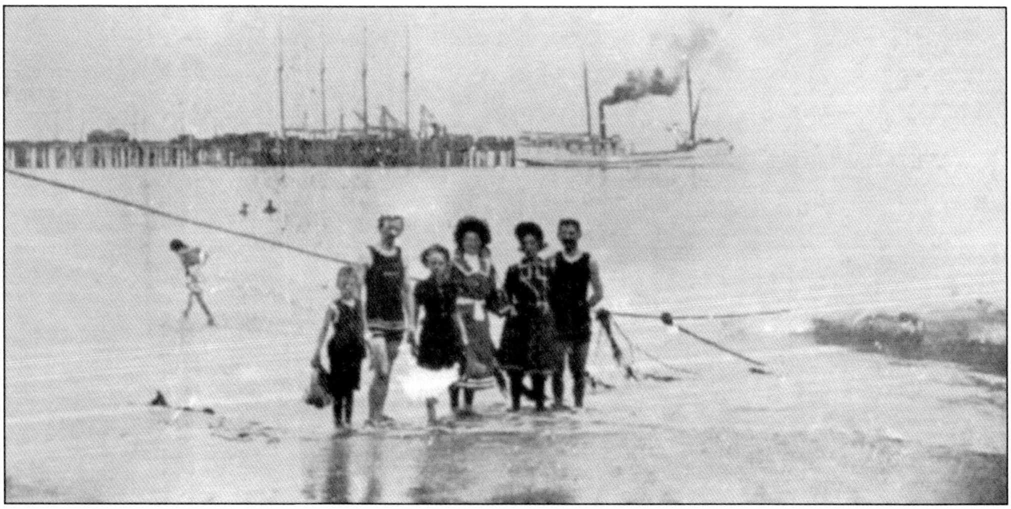

Bathing suits were more modest back then, and women even wore bonnets in the water. In this early-1900s photograph, the long rope behind the bathers was most likely tied to the Bath House and used by bathers as a safety line in case they were overpowered by the ocean waves.

Part of the appeal of the wharves, as well as the pleasure piers that came later, was the draw of the magnificent and mysterious sea. In the early 20th century, not many people knew how to swim, much less surf. In 1907, Henry Huntington met a young man named George Freeth, who was an avid swimmer and is thought of as California's first surfer. Surfing originated in Hawaii, and Freeth, being half Hawaiian himself, learned from the masters. Huntington hired Freeth as a lifeguard and showman for the new sport of surfing. He is a beloved character in the early beach resort history of Redondo Beach. At right is a photograph of Freeth. In the image below is one of the first lifeguard vehicles, designed by Freeth.

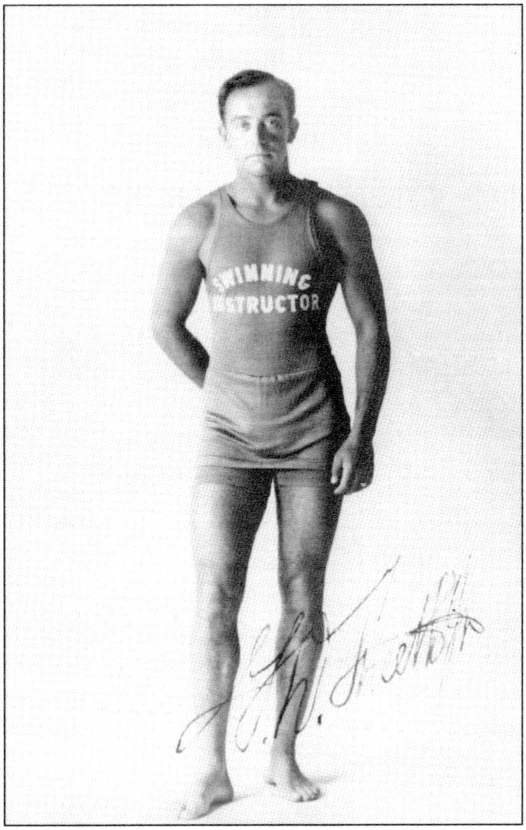

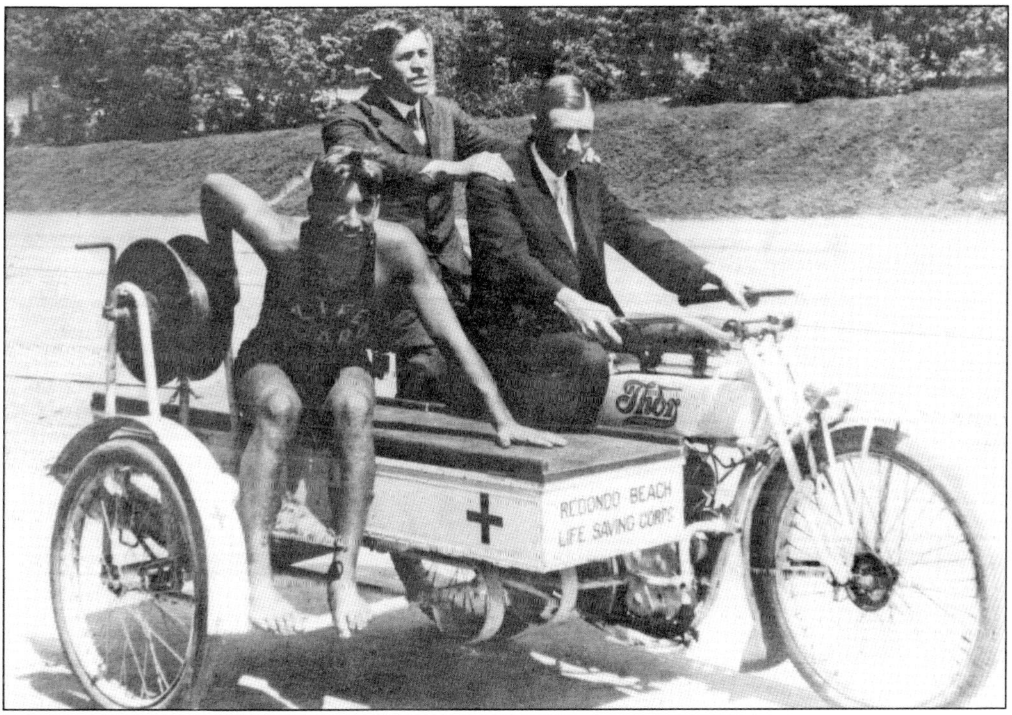

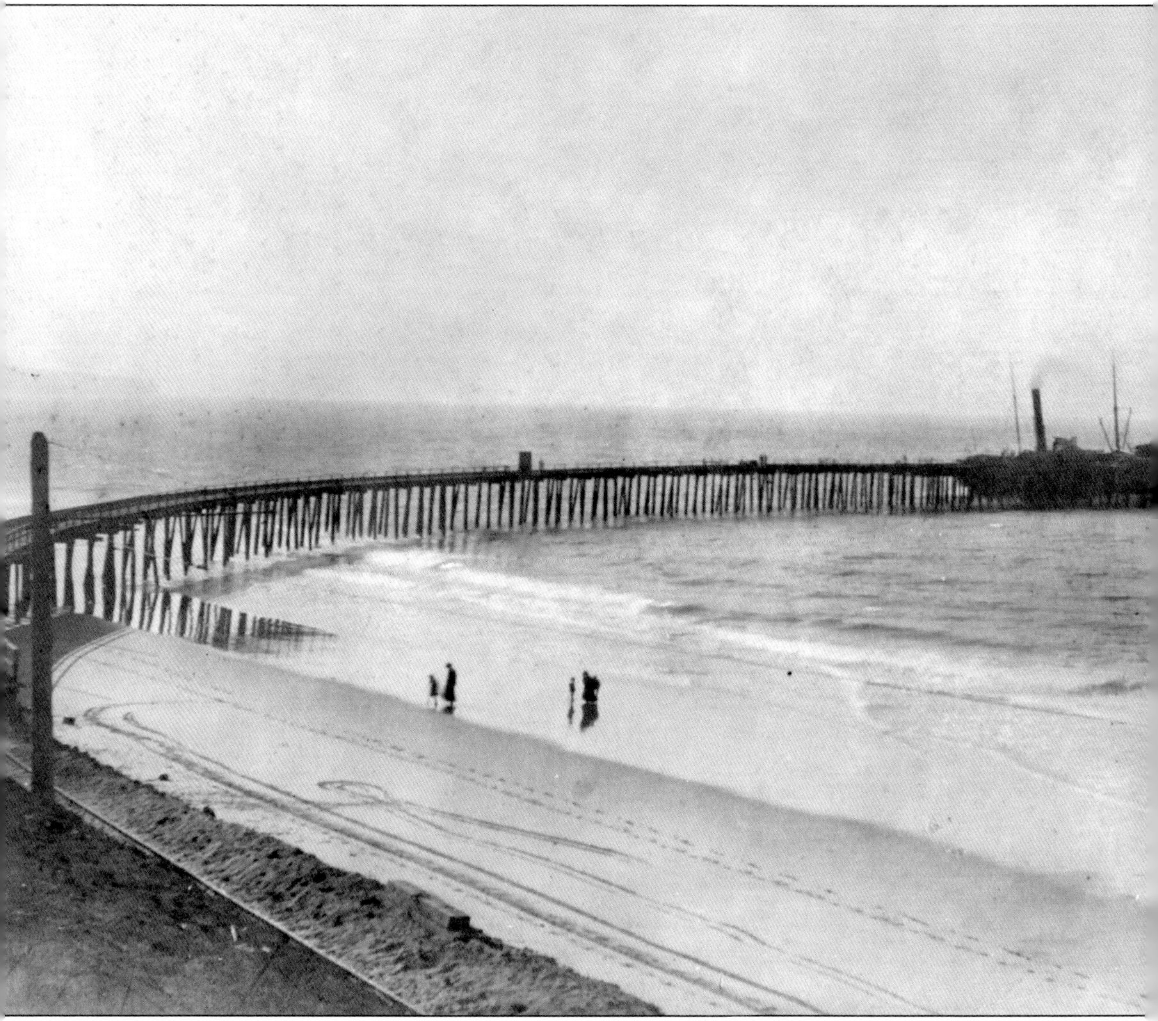

In 1904, Wharf No. 3 was constructed at the end of Sapphire Street, becoming the southernmost wharf. By the time it was built, San Pedro had become the main port for Los Angeles; however, Redondo Beach was still involved in the shipping business, with one of its main industries being lumber. In 1902, the railroad abandoned steam engines and switched to electric power. Real estate tycoon Henry Huntington—who is famous throughout the southland for his contributions to the Los Angeles area—started the Pacific Electric Red Cars operation around this time, but it did not reach Redondo for a few more years.

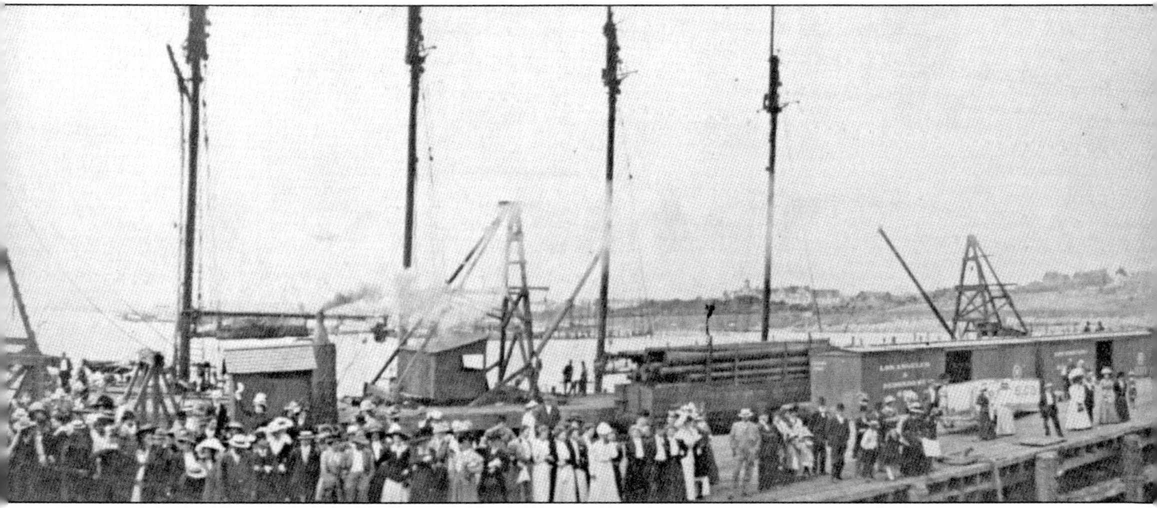

Passengers unload onto Wharf No. 3 at the end of Sapphire Street. In the background is the new town of Redondo Beach with only a few houses constructed on the bluffs overlooking the ocean. During this time in Southern California, beaches were a nice getaway for people who lived inland, but few actually resided permanently along the coast. By the early 1900s, more people set up permanent housing in beach towns as the tourism industry grew.

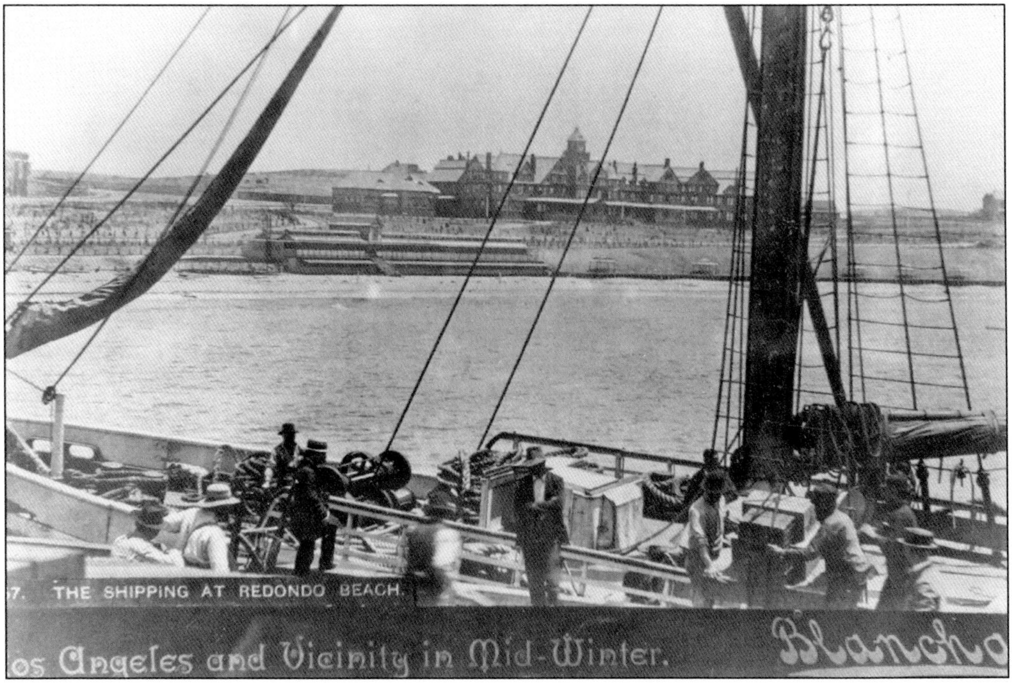

The shipping industry began at Redondo Beach in 1887, when the first railroad arrived and a deep water canyon was discovered off the coast of the shoreline. The Pacific Coast Steamship Company was one of the most lucrative operations utilizing Redondo's wharves. Alternately, the Santa Fe Railway also relied upon the shipping industry in the city and carried freight to Los Angeles, located roughly 20 miles away. The above photograph shows Redondo Beach in its early shipping port days. Hotel Redondo is in the background. Below is Redondo Beach years later with part of the new Pavilion built by Henry Huntington seen in the background.

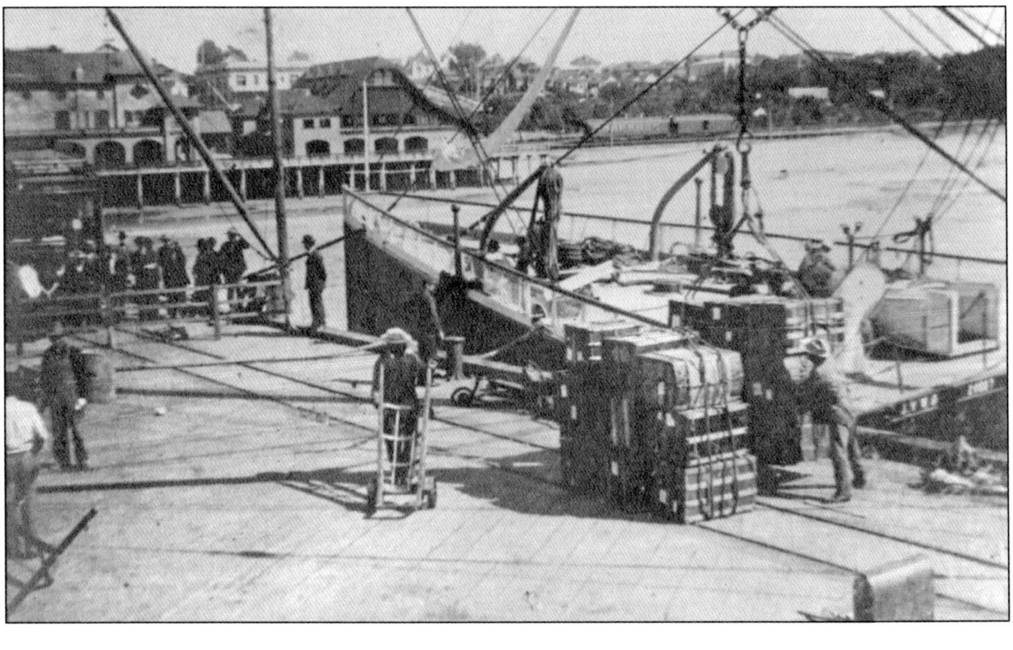

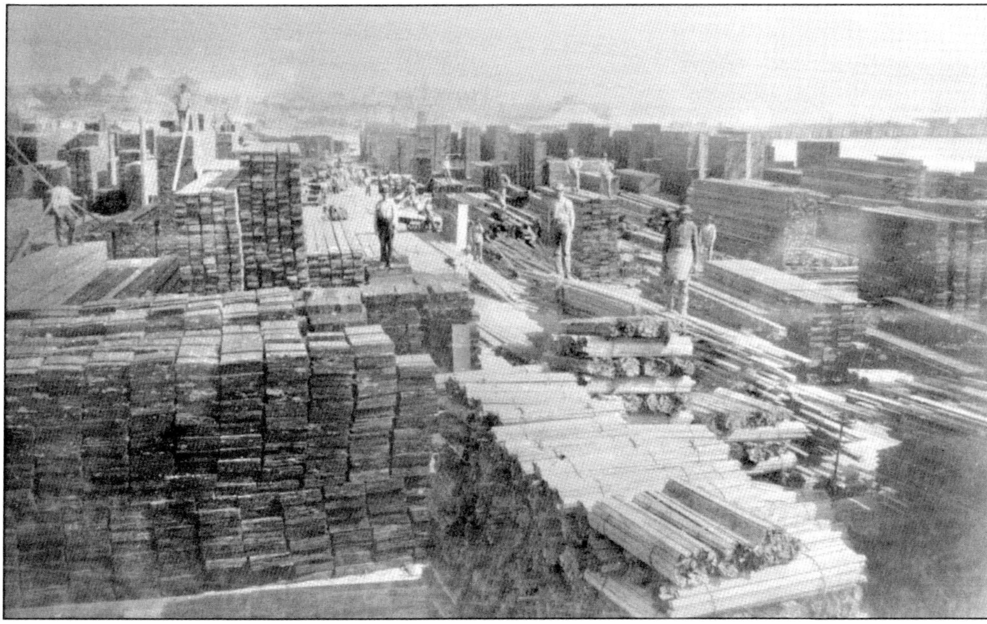

Lumber was a big industry in Redondo Beach. With an influx of people into Southern California, the demand for housing materials was high. In the late 19th century, more than half of virgin lumber came from the Pacific Northwest and was shipped to Redondo to be transported by train into the city of Los Angeles. There were two major lumber companies that developed in Redondo around 1904: the Ganhahl Lumber Company, which later became Redondo Lumber Company, and Montgomery & Mullen Lumber Company, which became Patten-Davies Lumber. Both firms had a planning mill operation at Broadway and Beryl Streets.

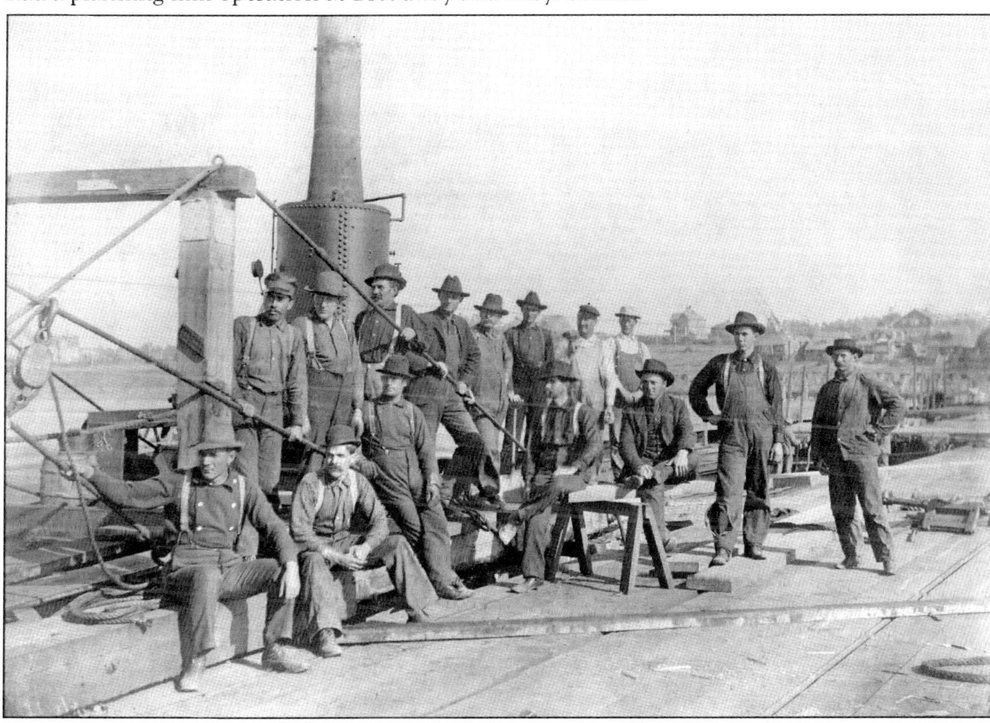

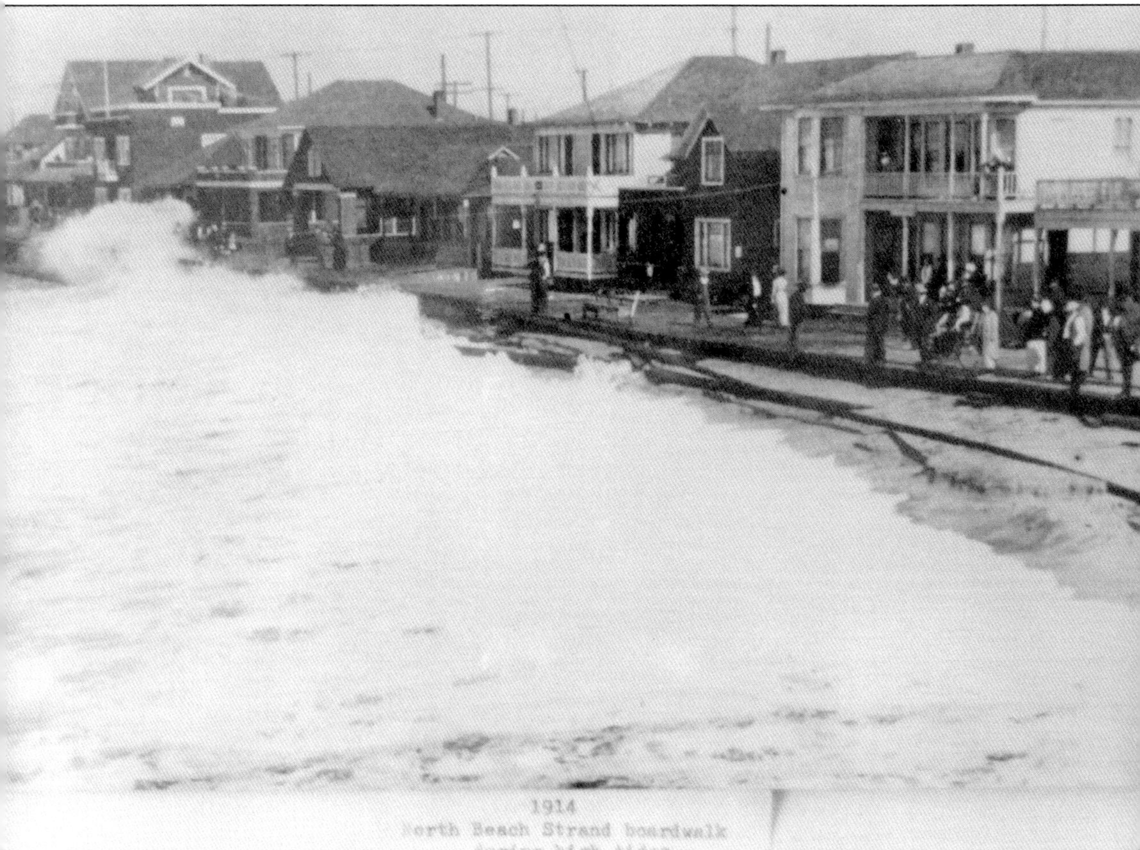

Storms often ravaged and destroyed the wharves of Redondo. The solution to this was a breakwater, an engineered structure used to reduce the intensity of waves and, therefore, prevent coastal erosion. However, a breakwater was not feasible to construct at Redondo because the wharves extended toward the underwater canyon. The idea of a floating breakwater was entertained but never constructed because no engineers had ever tried to build one yet. The first breakwater in Redondo was created in 1938–1939. It was necessary but was not constructed well enough for the future storms that continued to ravage and destroy the shoreline of Redondo Beach. In the mid-1950s, the breakwater was expanded and King Harbor was developed.

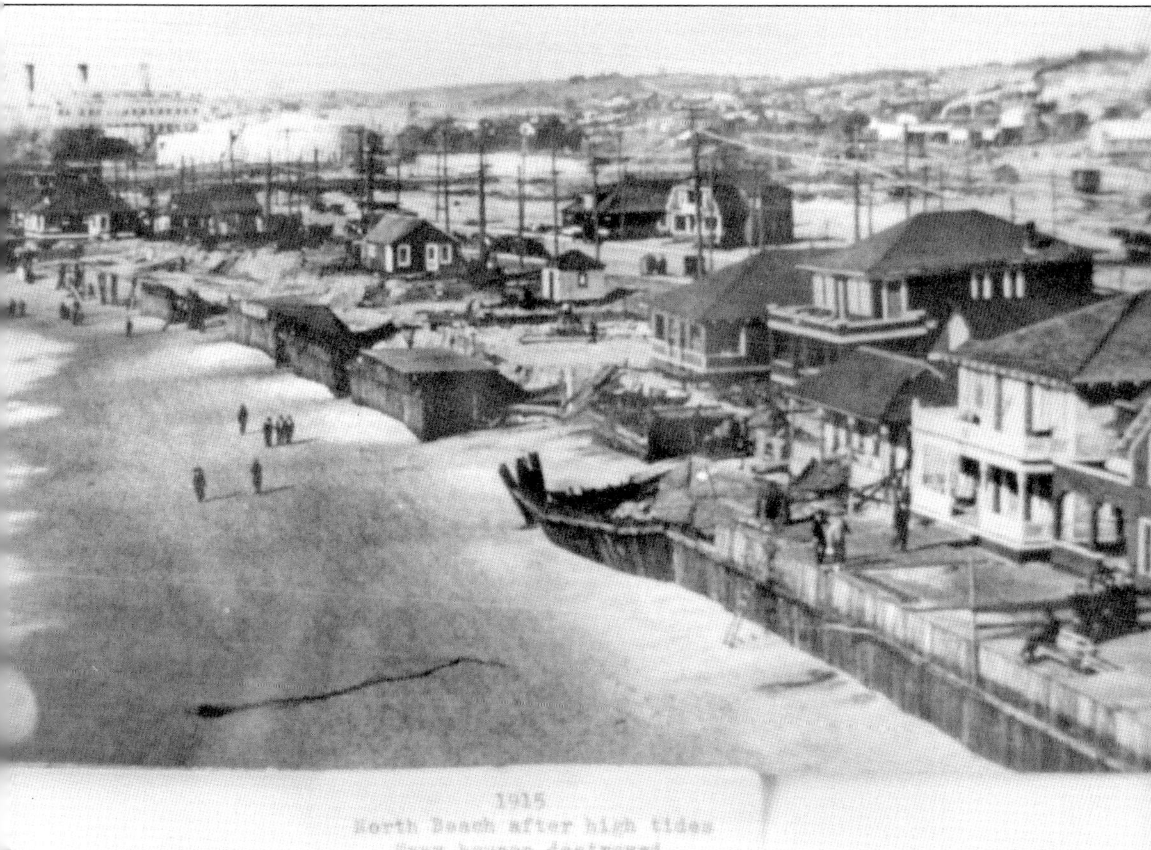

One of the worst storms in Redondo Beach history occurred in 1915. It destroyed wharves and swept homes into the ocean. Above is a photograph of one of the houses near the shoreline that was completely destroyed; the fence in front of it is gone.

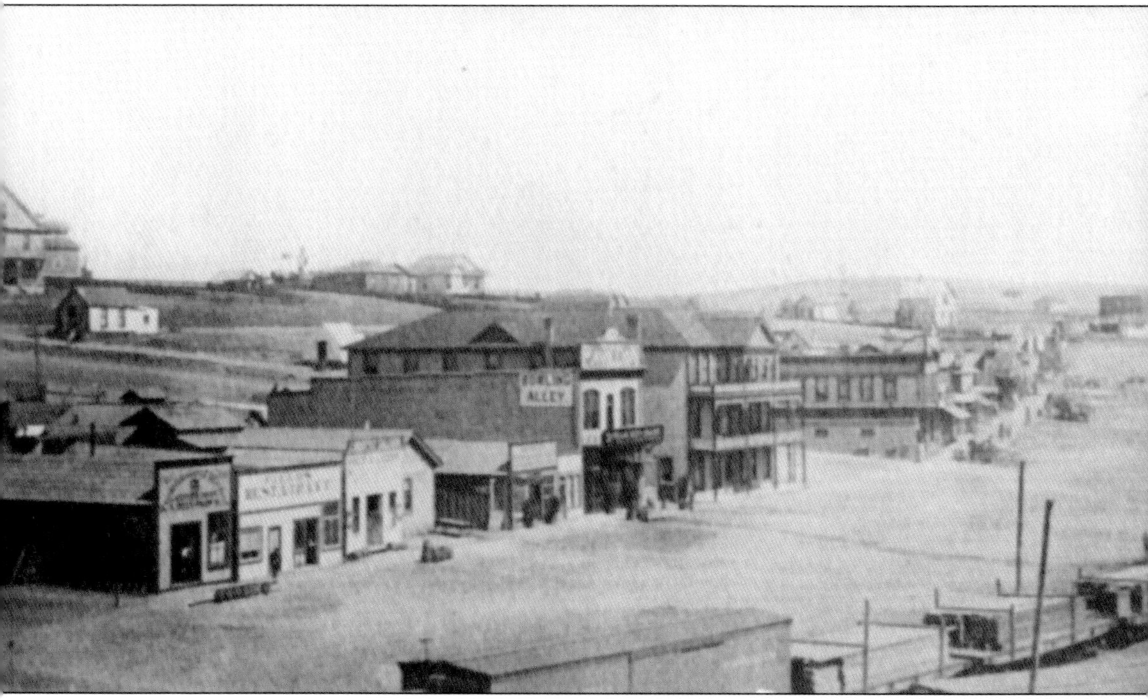

This is Redondo Beach as seen from the southeast on Pacific Avenue (now primarily condominiums) around 1890. The road is not paved, the storefronts have western false fronts with wood cladding, and there are only a few houses built on the hills in the background.

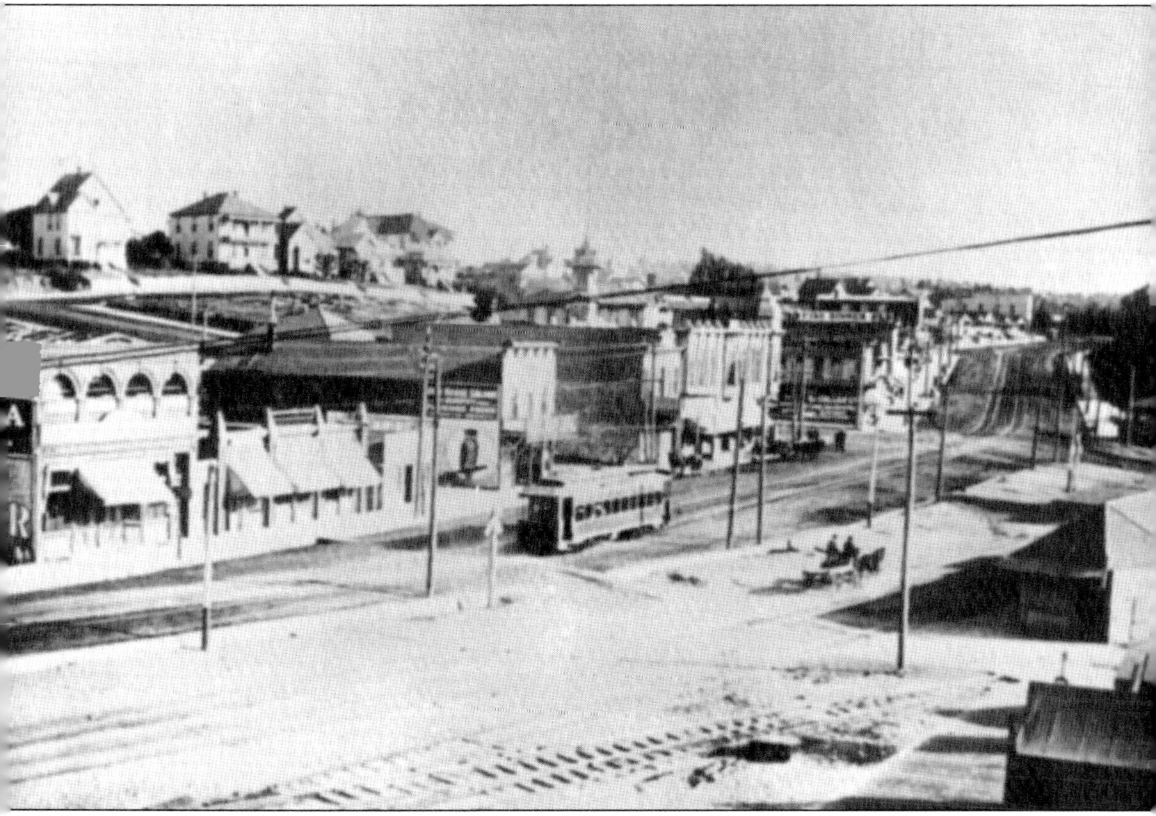

This photograph was taken some time after 1902 when electric trolleys were in service. Note that the street has tracks for the electric rail and is lined with poles for the electric wires that powered the trolleys. Also, there are many more buildings made of permanent materials such as brick. This shows the growth of Redondo in the early 1900s.

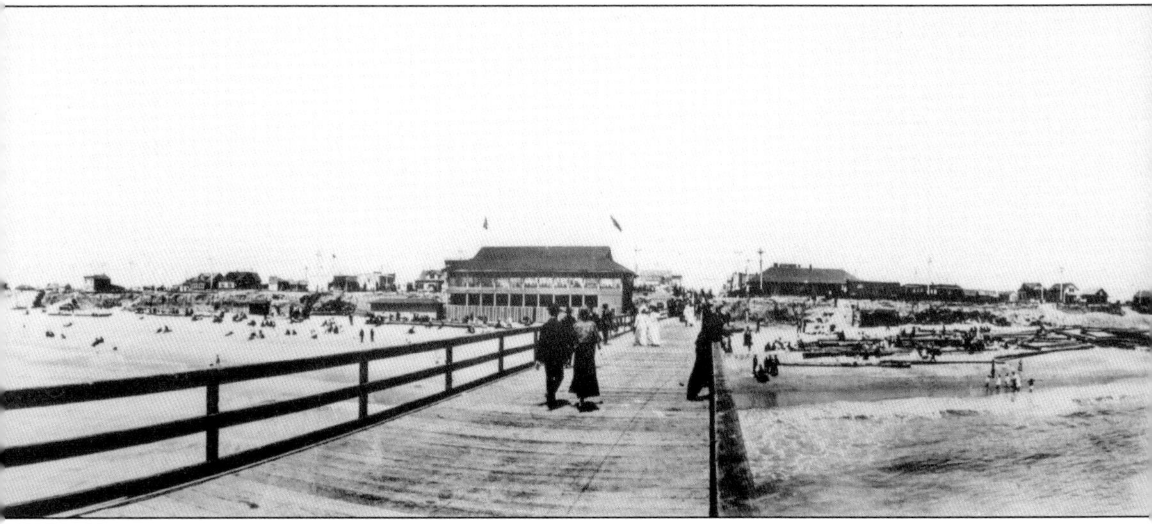

Pictured above is one of the Redondo Beach wharves in the early 1900s. Piers were usually located at the end of a prominent street. The first piers were typically constructed for utilitarian purposes. When business owners started to notice that piers attracted people, food and entertainment venues were constructed on the piers.

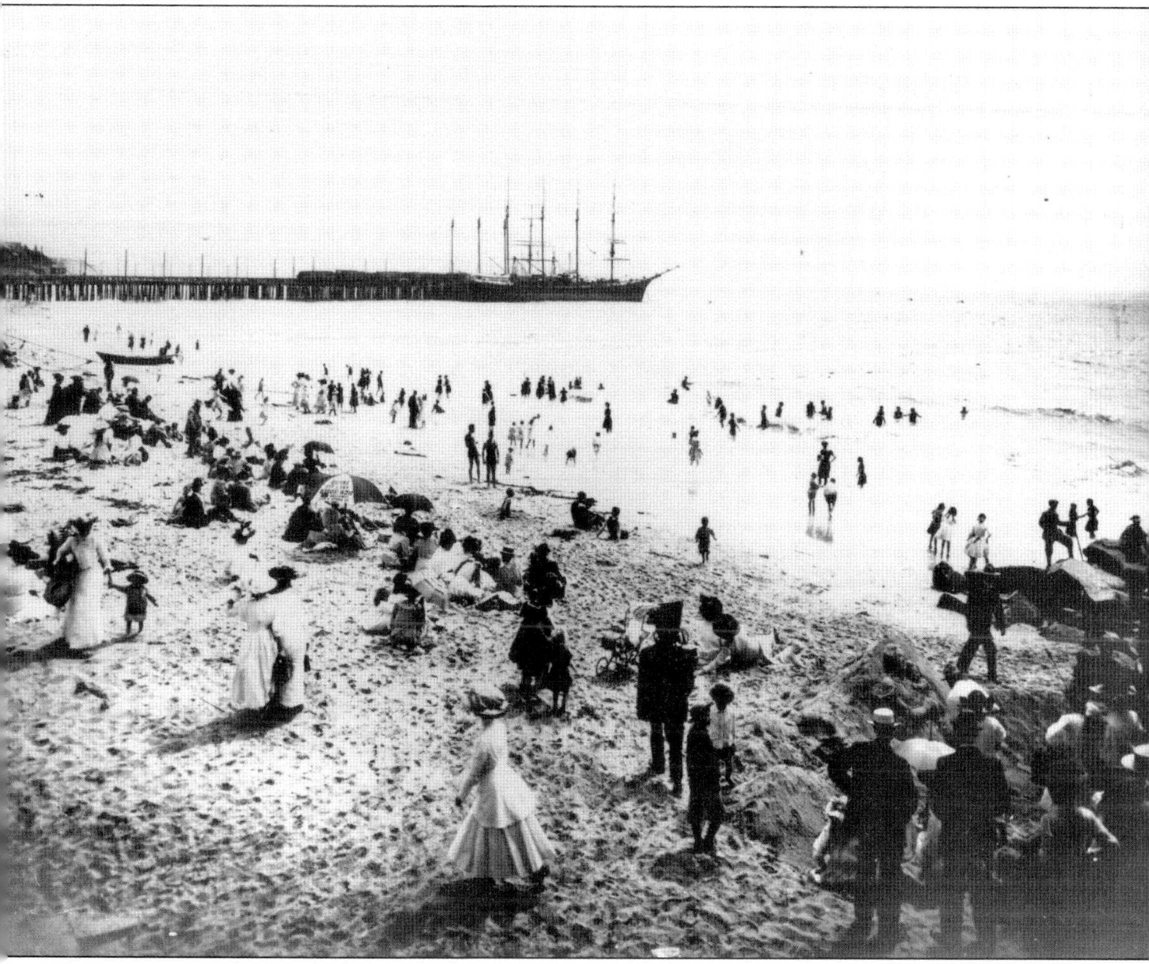

These are some of the early beachgoers in Redondo Beach in the late 1800s or early 1900s. Note the dress of the beach patrons. In the distance, a ship is at port at one of the wharves.

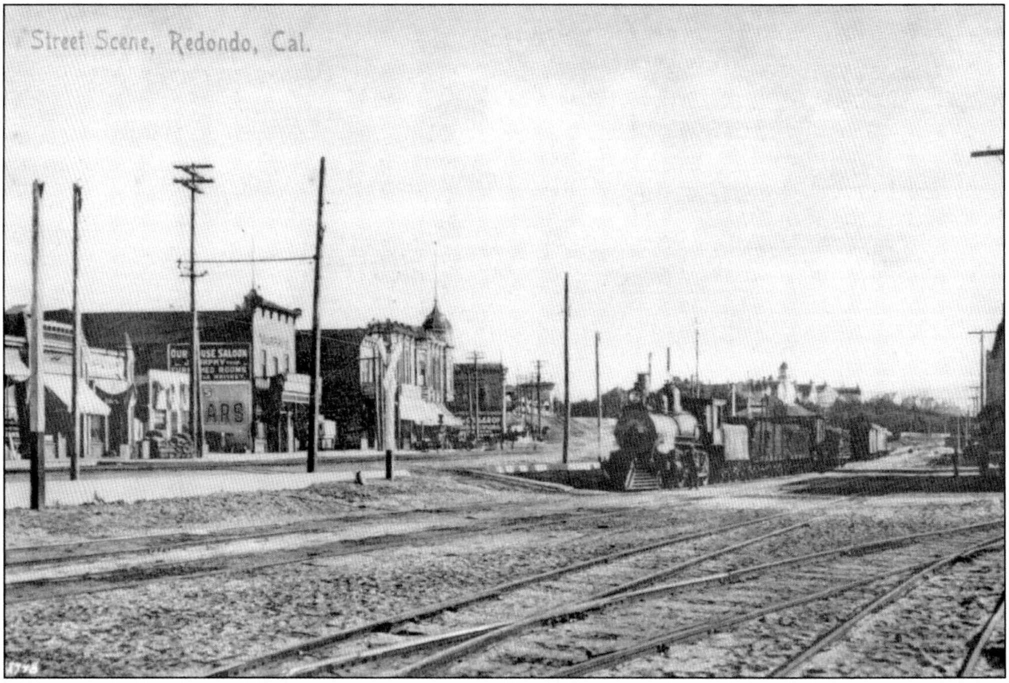

While Redondo was an important shipping port for Los Angeles, it was not seen as a major player in the Los Angeles Harbor war in the late 1890s. It was still a smaller operation compared to that of San Pedro and Santa Monica, but the shipping port days of Redondo Beach were integral to its development and eventual transformation into a beach resort town of the 20th century. Both of these photographs illustrate Redondo Beach in the early 1900s. The image above is located near where Pacific Avenue and the later El Paseo split just north of Hotel Redondo. Below is Emerald Street where it once met Wharf No. 1.

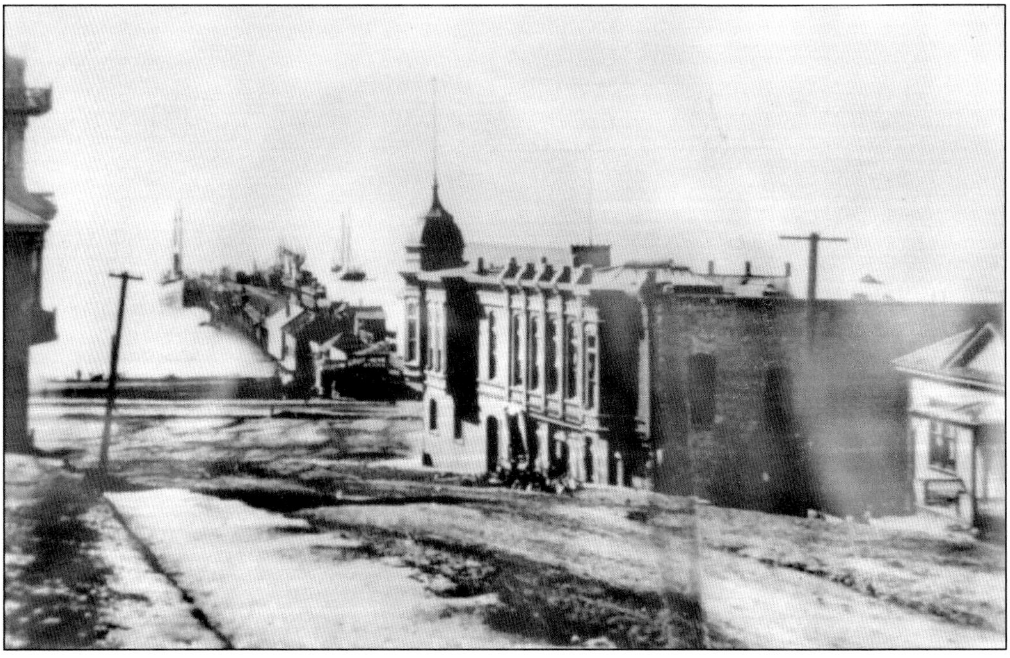

Two

The Beach Resort and the Redondo Beach Pleasure Piers

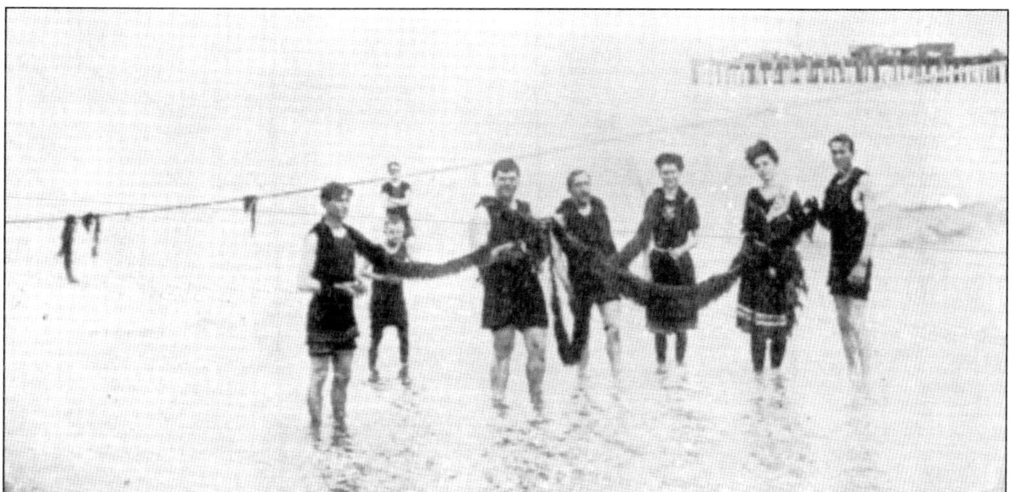

Beginning in 1905, Henry Huntington took over the Redondo Beach Improvement Company and bought out the Los Angeles & Redondo Railway. He replaced the engines with his own Pacific Electric Red Cars. His vision for Redondo Beach was not focused on Redondo's potential as a shipping port but instead as a beautiful beach resort that catered to tourists.

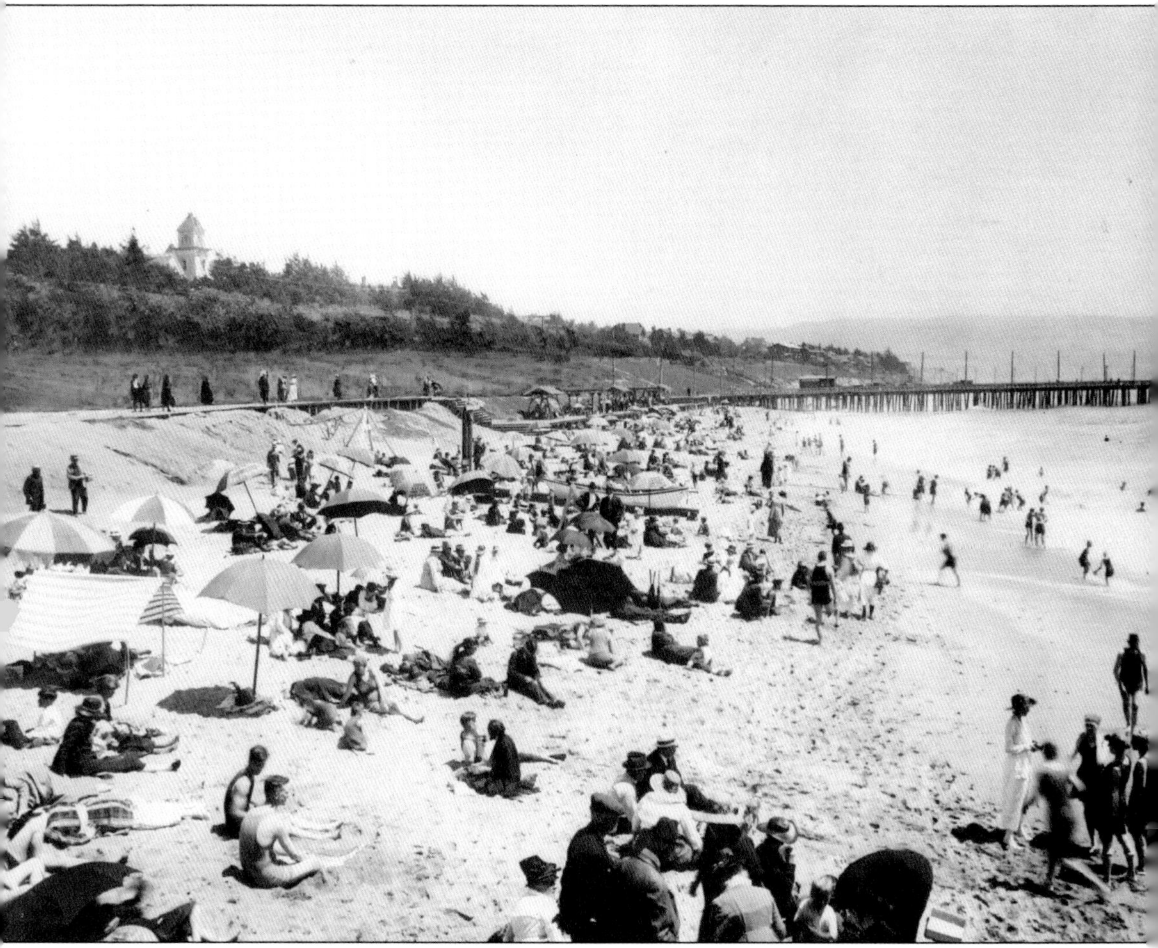

Because San Pedro had recently received government money for improvements and updates to its harbor, it became the major port for Los Angeles by the 1890s. While Redondo Beach was still involved in the shipping industry, it began to draw more tourists. Henry Huntington's Pacific Electric Red Cars began running in Redondo some time around 1905, which made it easier for inlanders from as far as San Bernardino to access Redondo Beach. The Pacific Electric Red Cars began to bring hundreds to thousands of tourists to the beach daily.

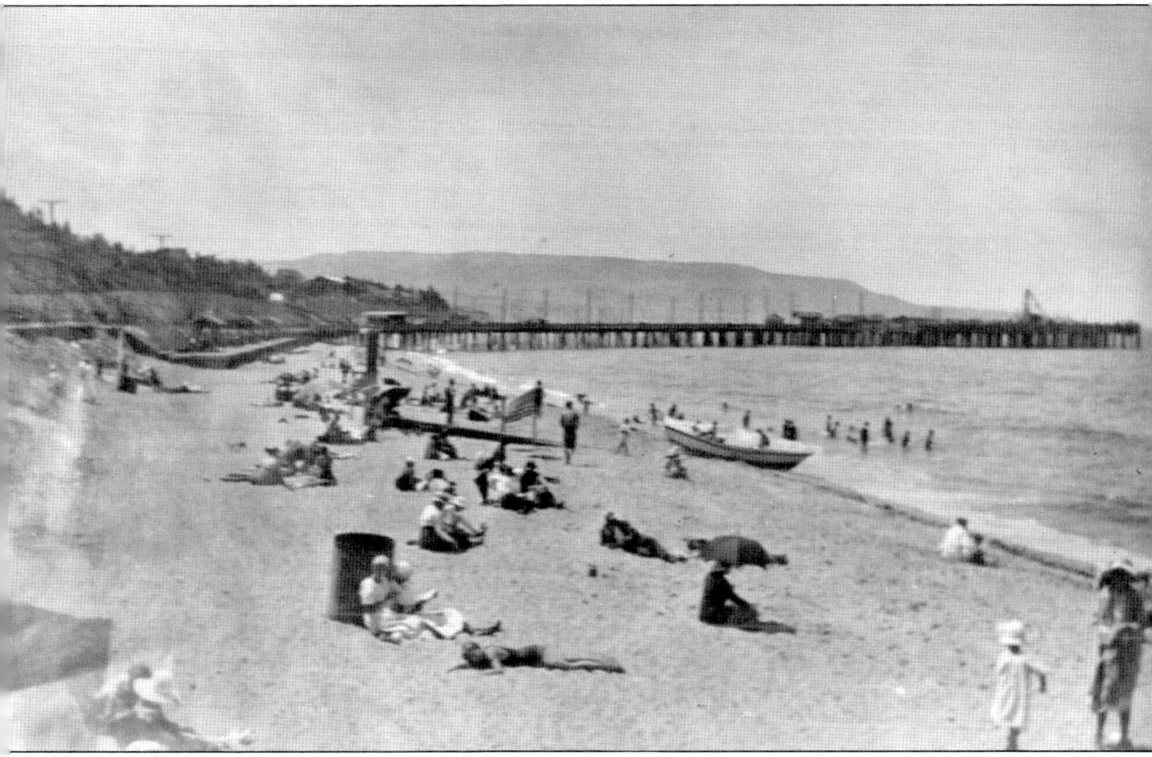

Pictured above are some beachgoers enjoying Redondo Beach. This photograph was taken from the north looking south towards the Palos Verdes Peninsula and Wharf No. 2.

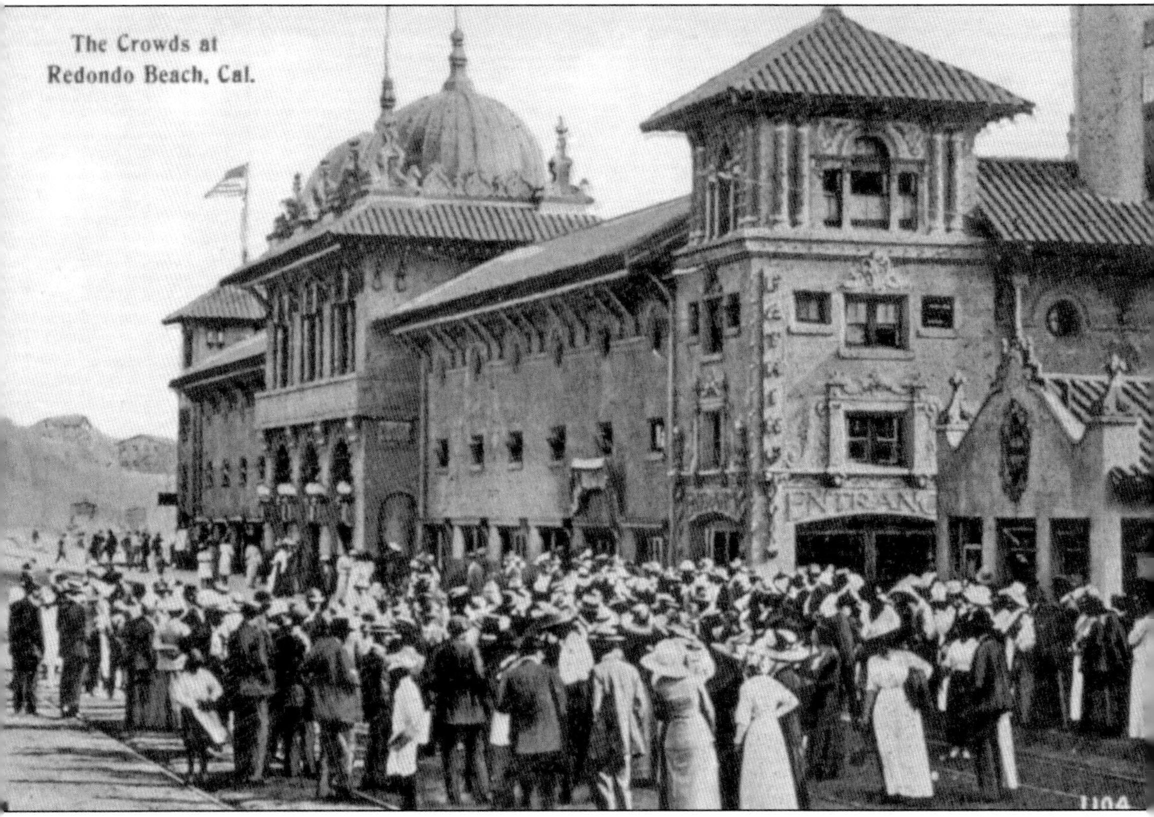

One of the first things Henry Huntington did to promote Redondo Beach's seaside was build a pavilion. This postcard is from around 1907, after the Redondo Pavilion was constructed to replace the outdated original pavilion. It featured the Mandarin Ballroom, which seated up to 4,000 people. The ballroom was on the second story, while shops, a theater, and a restaurant occupied the first floor. (Courtesy of Donald Krintz.)

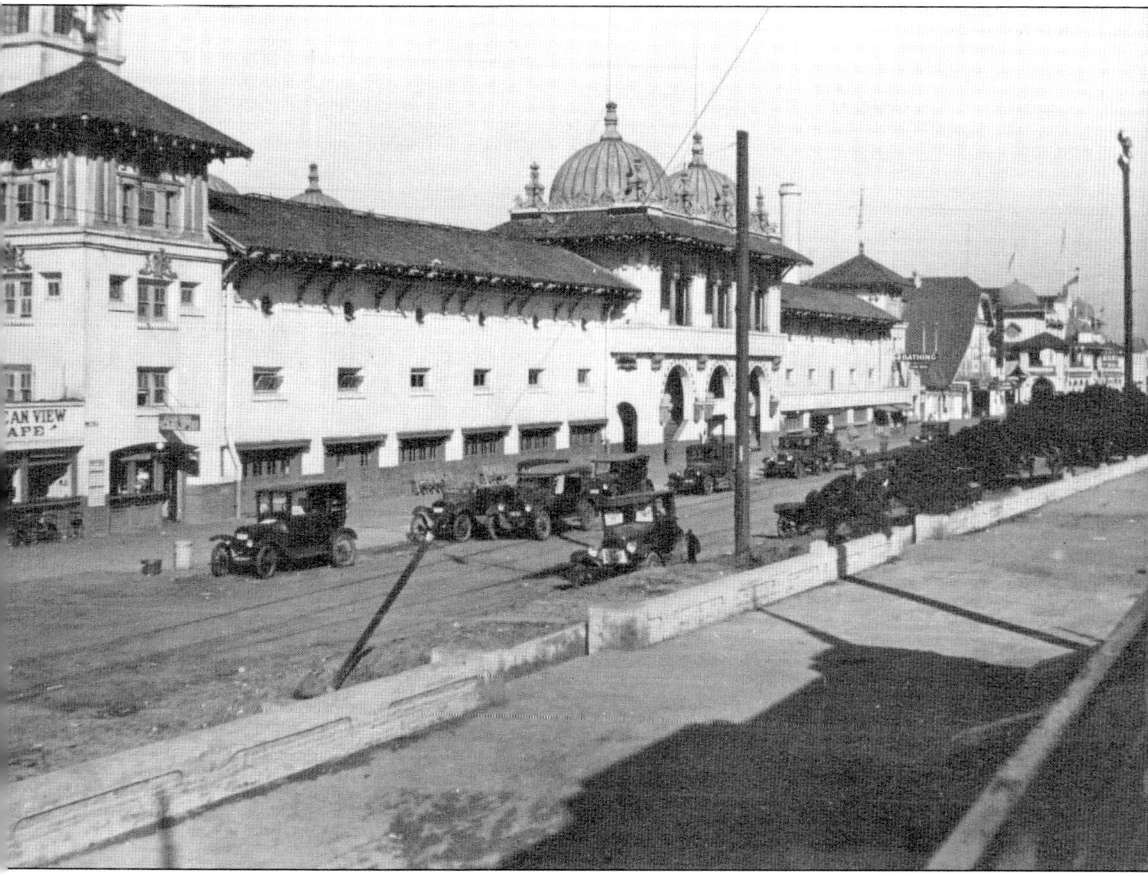

The second extravagant building constructed under the direction of Henry Huntington was the Bath House, also called "the Plunge." This building was constructed shortly after and just north of the Pavilion in 1908. Both buildings replaced the older, smaller wooden structures of the late 19th century. Their architectural grandeur was celebrated and enticed tourists from all over Southern California to come to Redondo Beach.

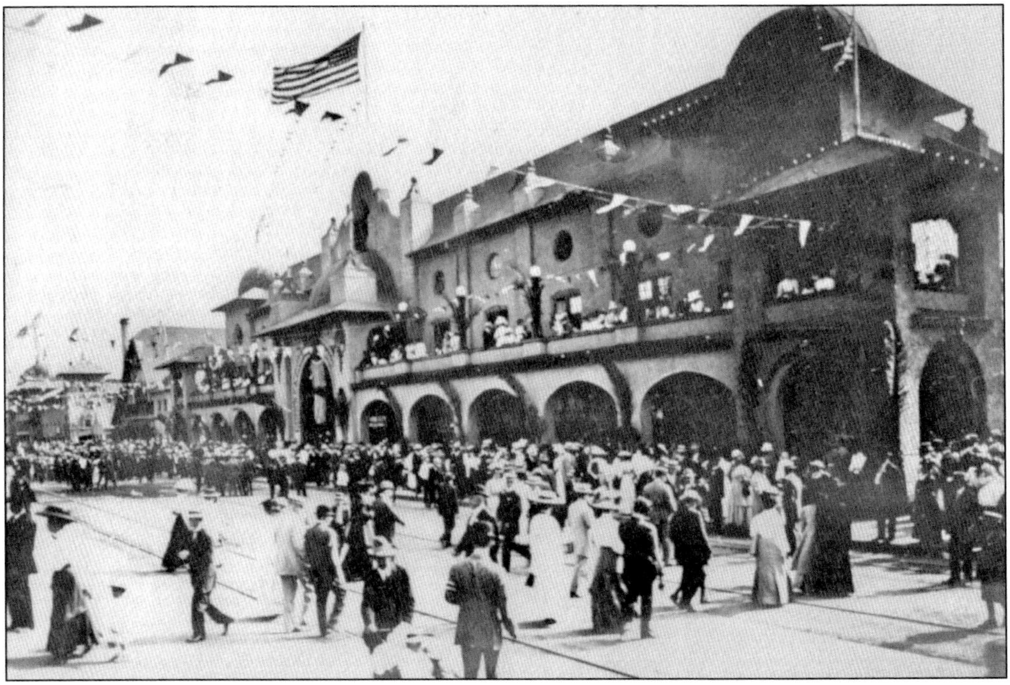

The Redondo Pavilion and the Bath House were the main architectural features on a famous thoroughfare of the early 1900s in Redondo Beach. El Paseo, the street nearest to the beach, was where Huntington's Pacific Electric Red Cars dropped off and picked up hundreds of tourists daily. El Paseo had several tourist shops, entertainment, and amusements. Its close proximity to the seaside made a day at the beach a fantasyland. The area between the Pavilion and the Plunge was called the Midway for some time. Exotic architecture, flags, decorative signage, and elaborate ornamentation on the buildings added to the excitement of being at the beach.

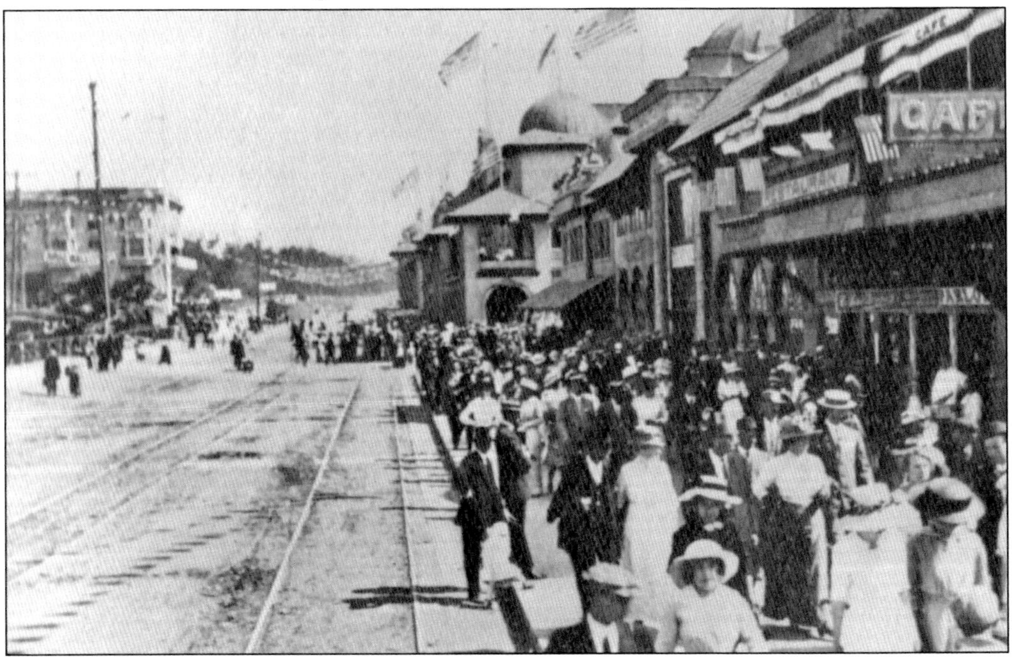

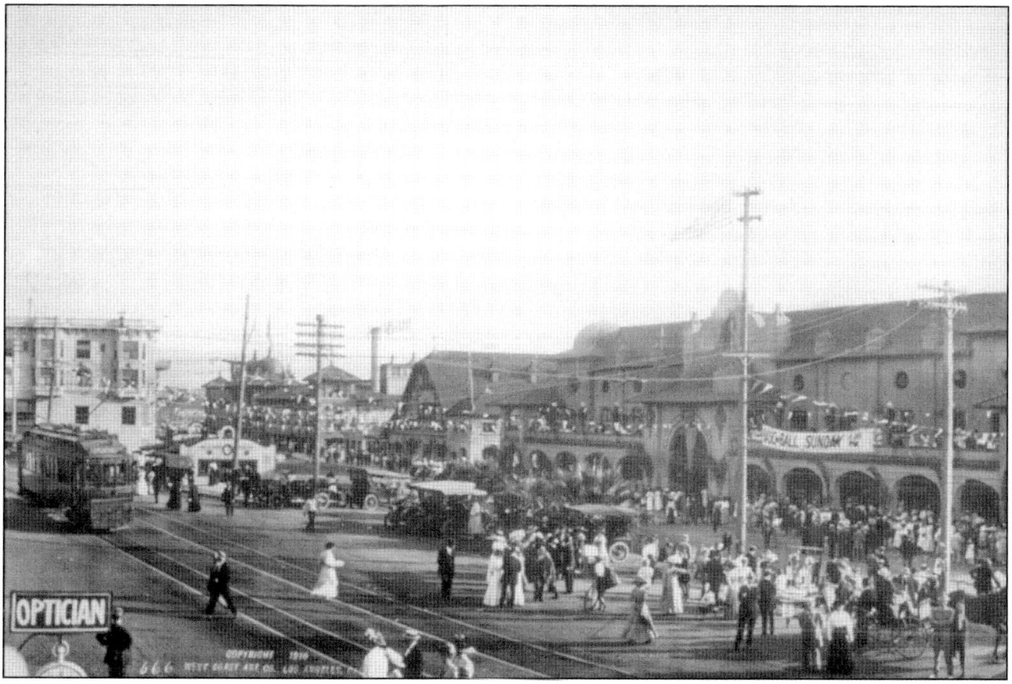

This early-1900s view of El Paseo was taken from the northeast. In the right foreground is the rear of the Redondo Pavilion, and to the far left is the Bath House. Also to the left is one of the Pacific Electric Red Cars riding down the track on Pacific Avenue.

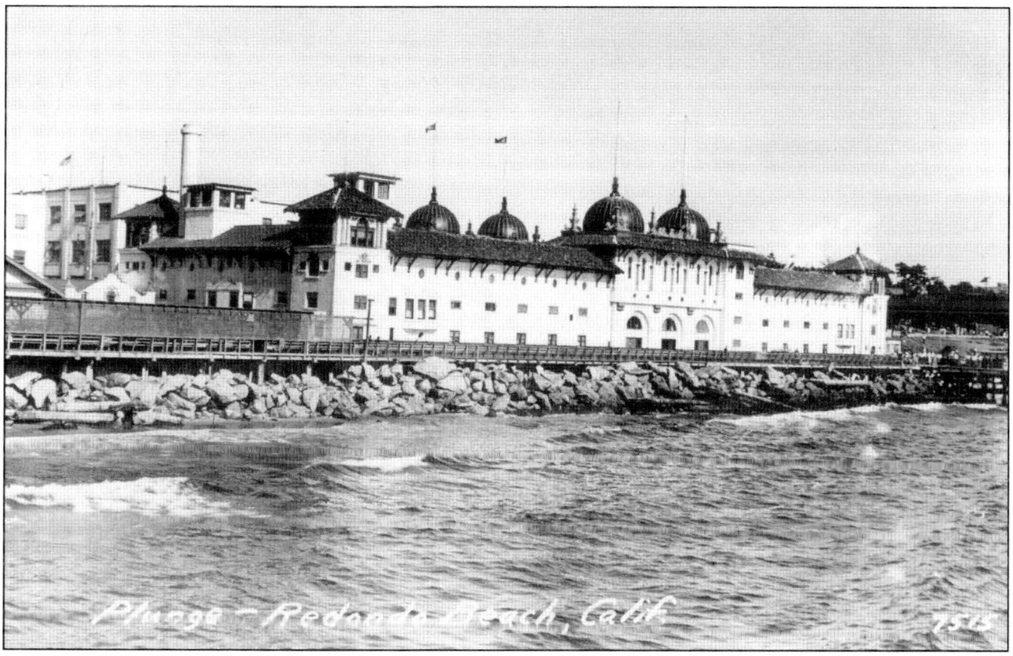

This photograph of the Bath House, also referred to as "the Plunge," was taken from the Endless Pier sometime around 1916. It was 43,688 square feet and known as "the largest indoor saltwater pool in the world" during the early 1900s. It featured three heated saltwater swimming pools that could accommodate approximately 2,000 people.

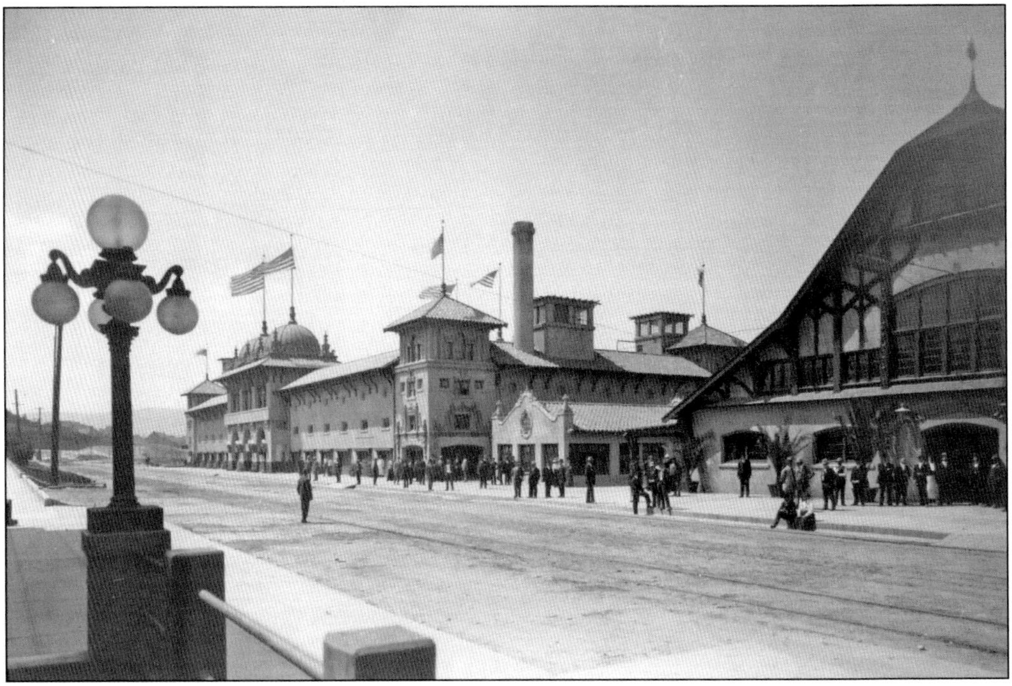

El Paseo was the main attraction before the Endless Pier was built. Here is a picture of a quiet day on El Paseo. In the right foreground is the casino, located north of the Pavilion. The beach is hidden behind these structures.

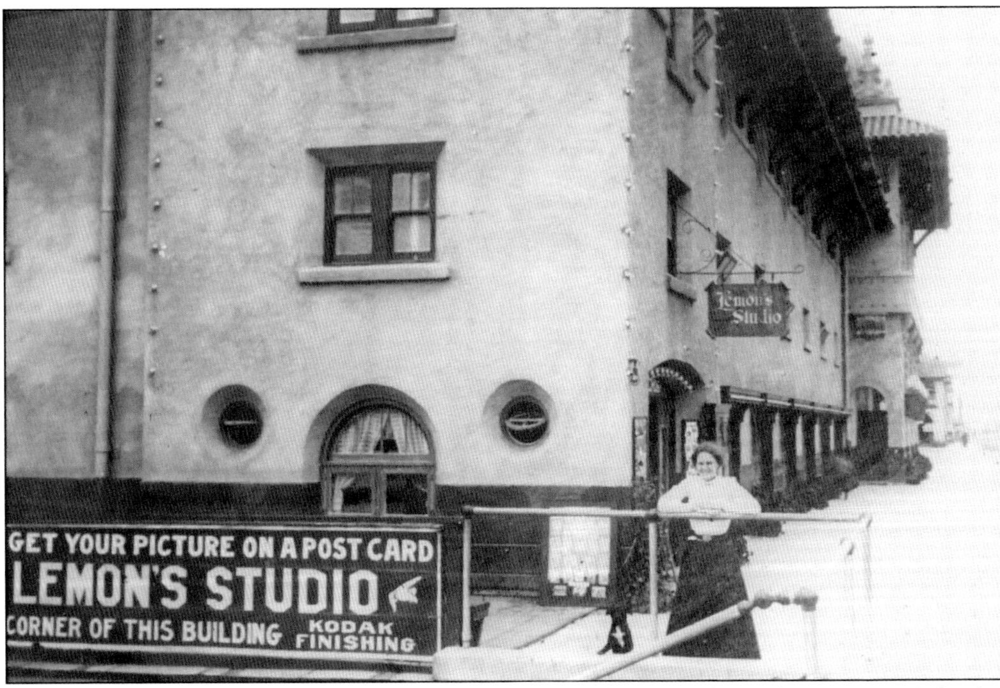

Pictured around 1907 is the Redondo Pavilion and one of the featured storefronts, Lemon's Studio. The sign reads, "Get Your Picture on a Post Card, Lemons' Studio, Corner of this Building." The studio also offered "Kodak Finishing."

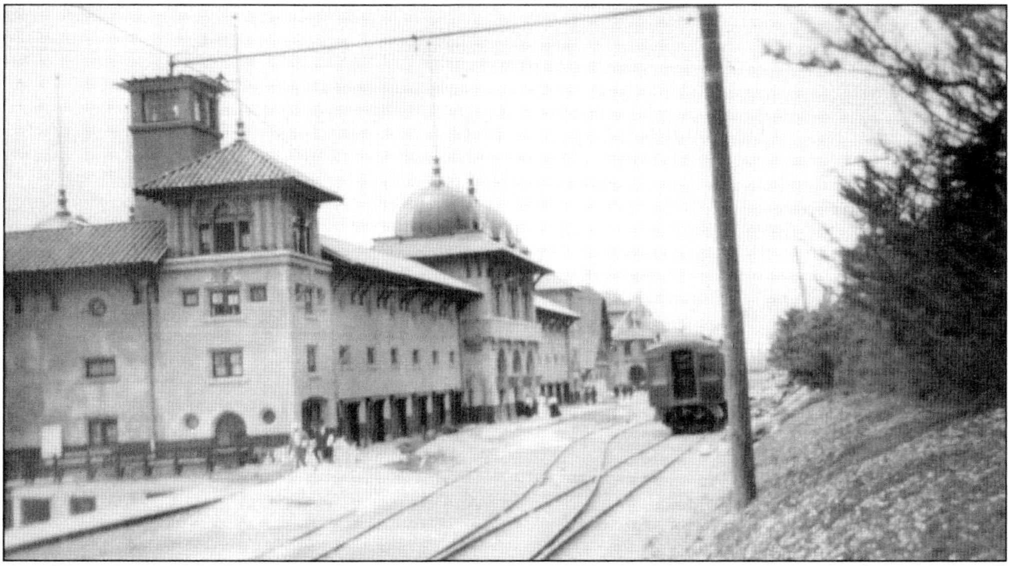

Henry Huntington's Pacific Electric Red Cars were a popular mode of transportation but not just in Redondo Beach. It was also the main interurban transportation route in Southern California from the late 19th century up until World War II. The Pacific Electric Red Cars company was incorporated in 1901, and tracks were immediately laid through Los Angeles, Orange, San Bernardino, and Riverside Counties.

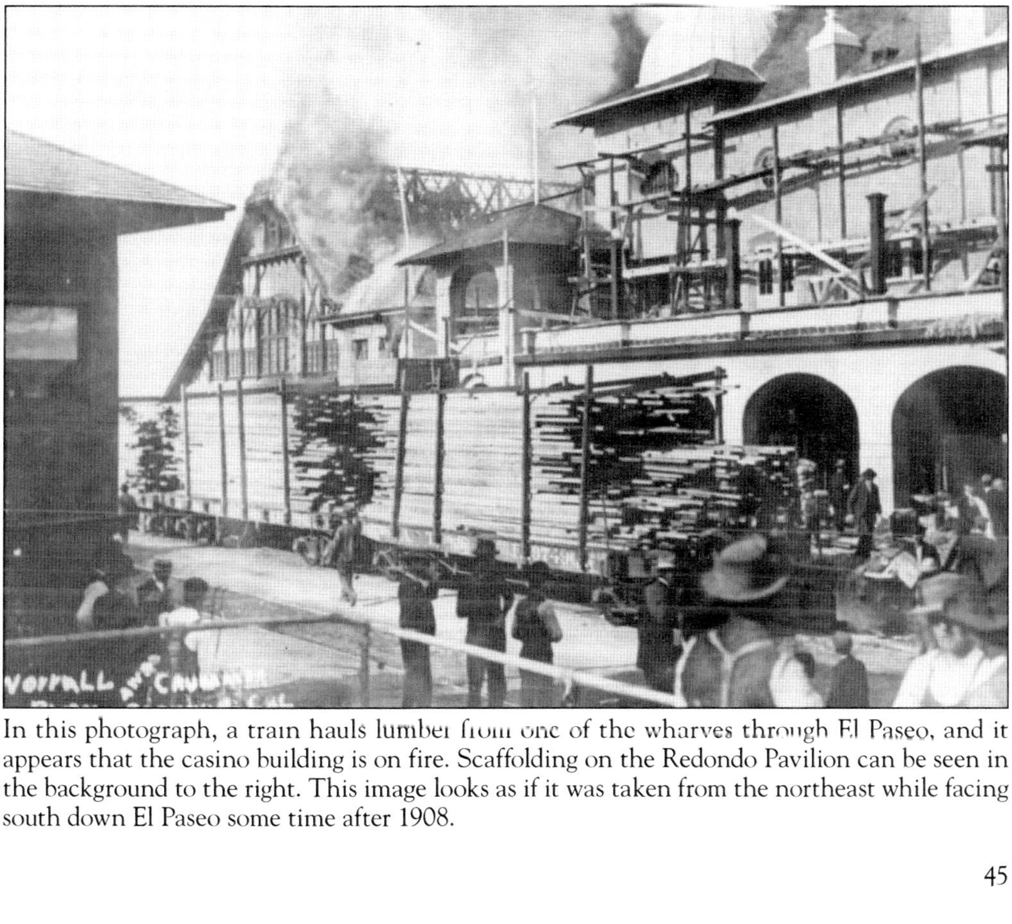

In this photograph, a train hauls lumber from one of the wharves through El Paseo, and it appears that the casino building is on fire. Scaffolding on the Redondo Pavilion can be seen in the background to the right. This image looks as if it was taken from the northeast while facing south down El Paseo some time after 1908.

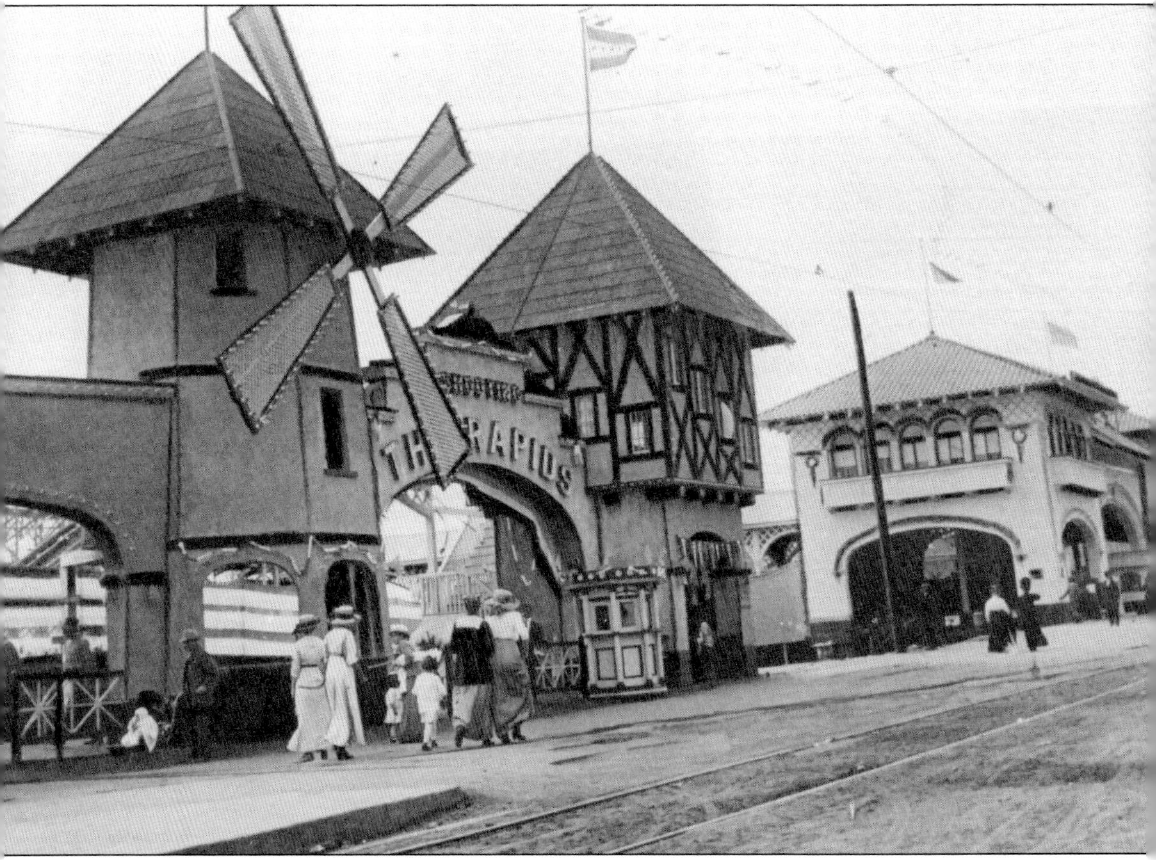

This is another view of El Paseo from around 1910. The grand entryway with the large windmill seems to be the entrance to the amusement zone, which was located north of the Pavilion on El Paseo. Behind this entrance, which can barely be seen, is the famous Redondo Beach Lightning Racer roller coaster.

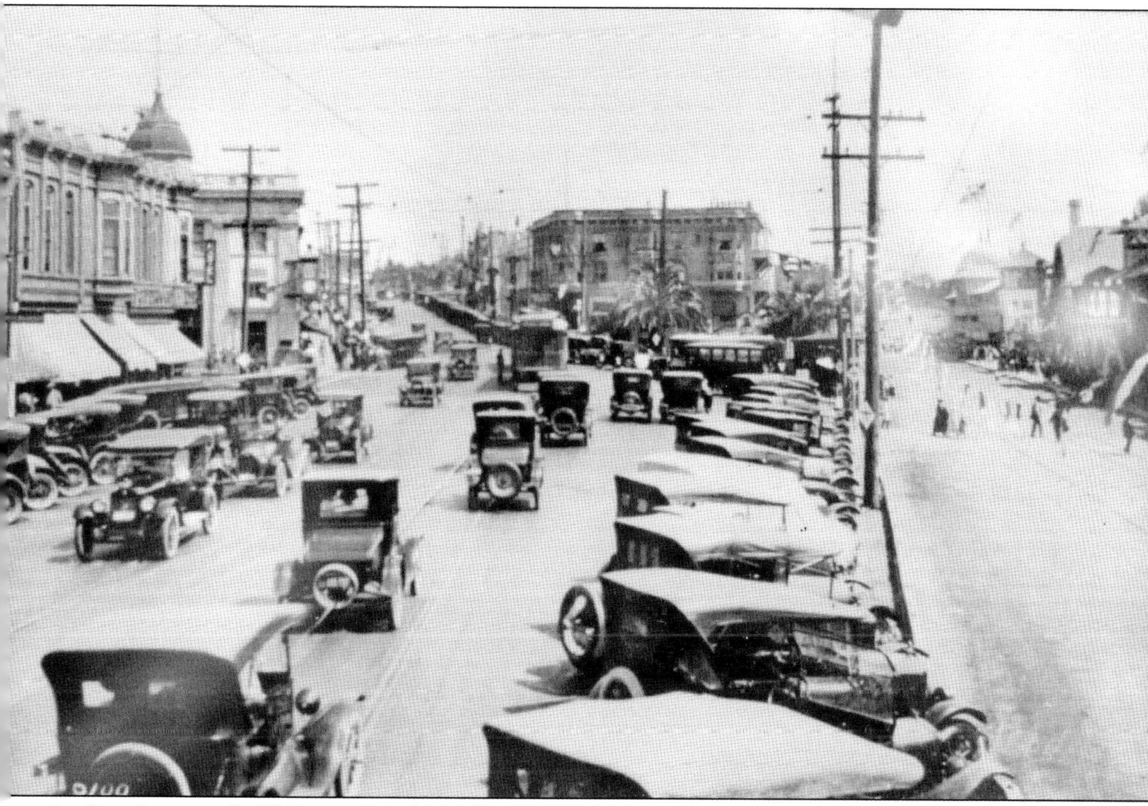

In this photograph, El Paseo is to the right and Pacific Avenue is to the left. Today, this area has been redeveloped with large condominiums. Catalina Avenue and the Esplanade are now the main thoroughfares beside the beachfront.

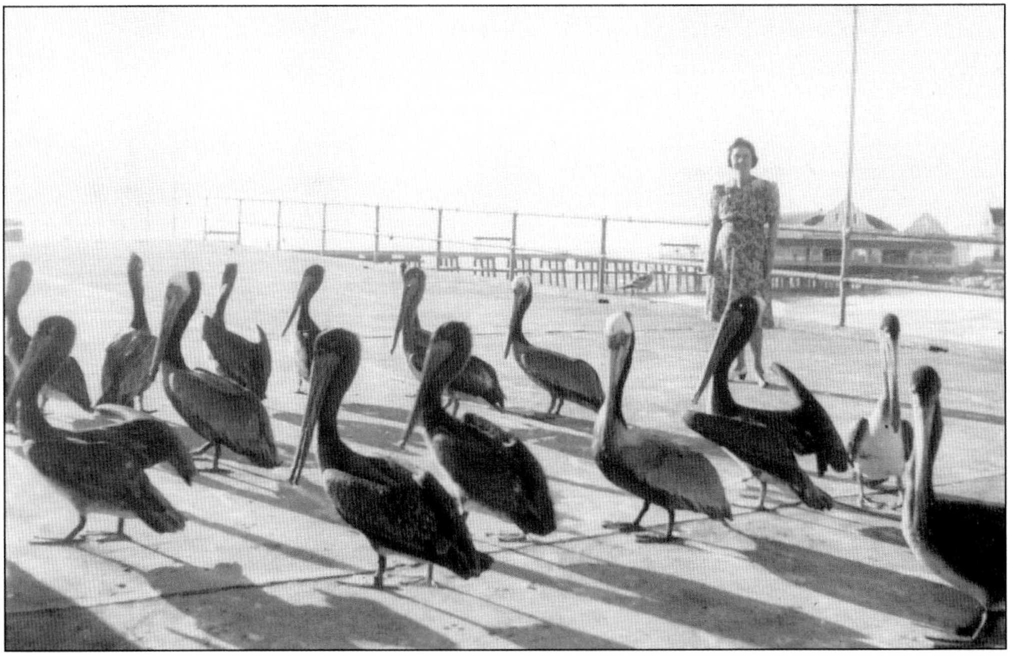

Pelicans parade down the pier as a woman poses for a photograph behind them.

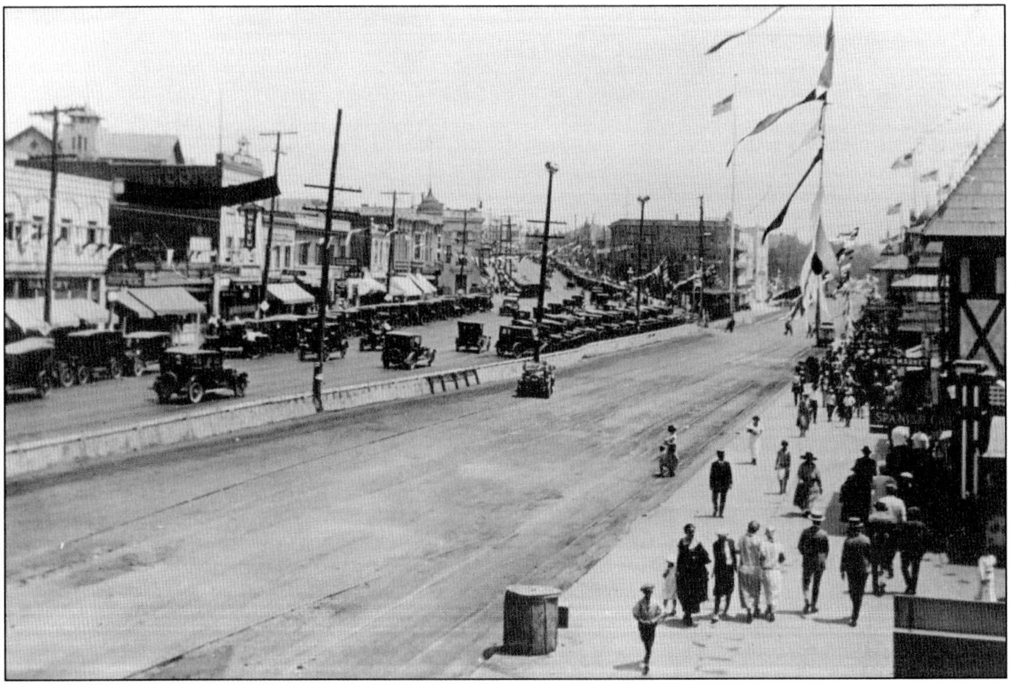

This image is of El Paseo and Pacific Avenue in the 1920s. Some of the entertainment venues featured along El Paseo included penny arcades, dodgem cars, the Hippodrome merry-go-round, a tunnel of love, a fun house, shooting galleries, games, restaurants, and boardwalk cafés. In the background is Pacific Avenue. A 1911 article in the *Los Angeles Times* reported that during one weekend in July there were "one hundred and ten automobiles" lining Pacific Avenue, the most cars ever reported in Redondo Beach at the time.

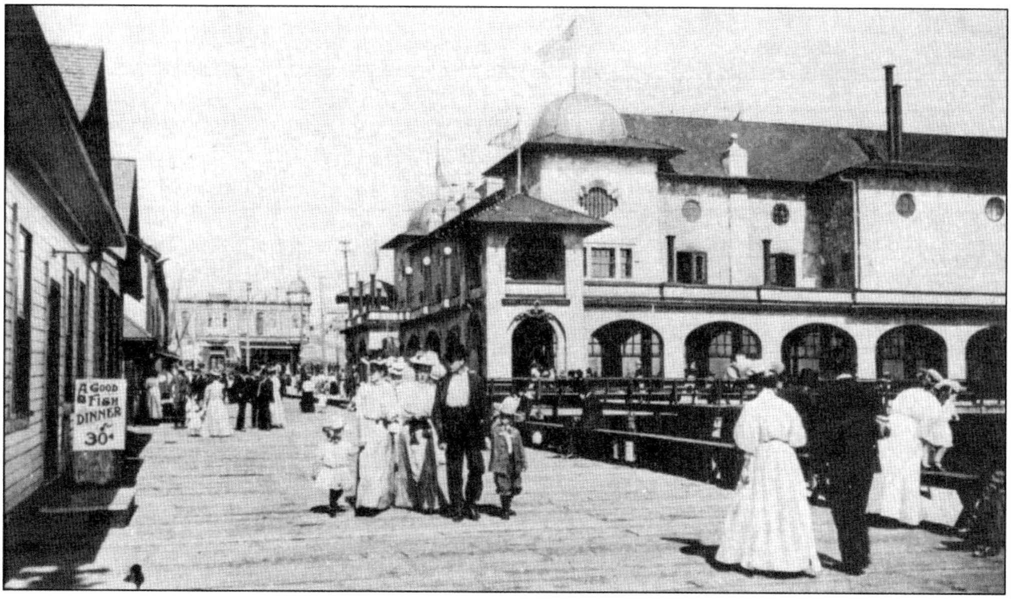

In 1911, about 300 men and women in Hermosa and Redondo Beach signed a petition opposing dancing at the Pavilion on Sundays. These beach resort towns are located in close proximity to one another, and according to a *Los Angeles Times* article, a group of protestors felt strongly that dancing on Sundays "will be detrimental to both places." However, it was not determined if this petition ever became a law. Above is a picture of the Pavilion and its visitors enjoying a stroll to and from the pier. In the distance, Emerald Street can be seen. The image below, taken around the same time period, is a postcard with a view of the Plunge and the pier from the north. (Below, courtesy of Donald Krintz.)

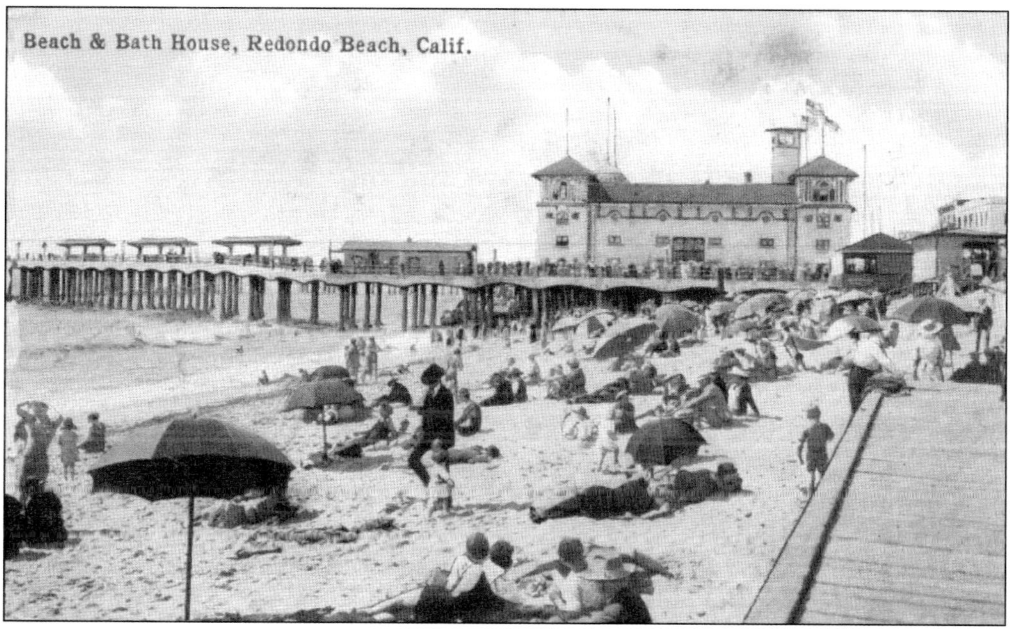

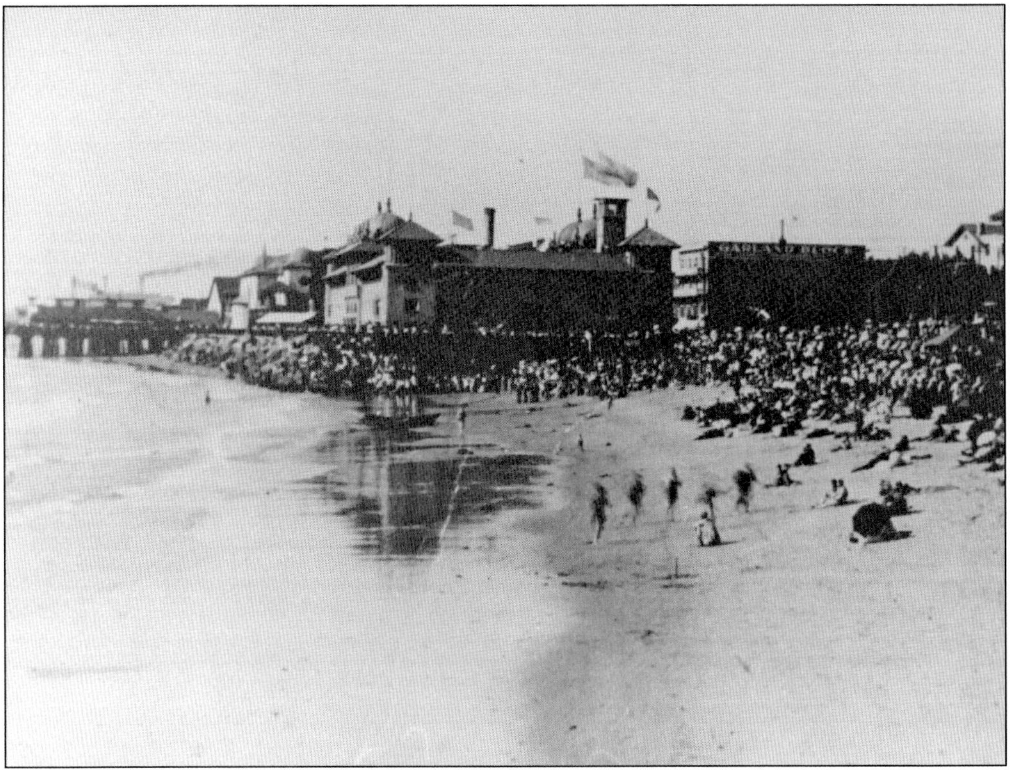

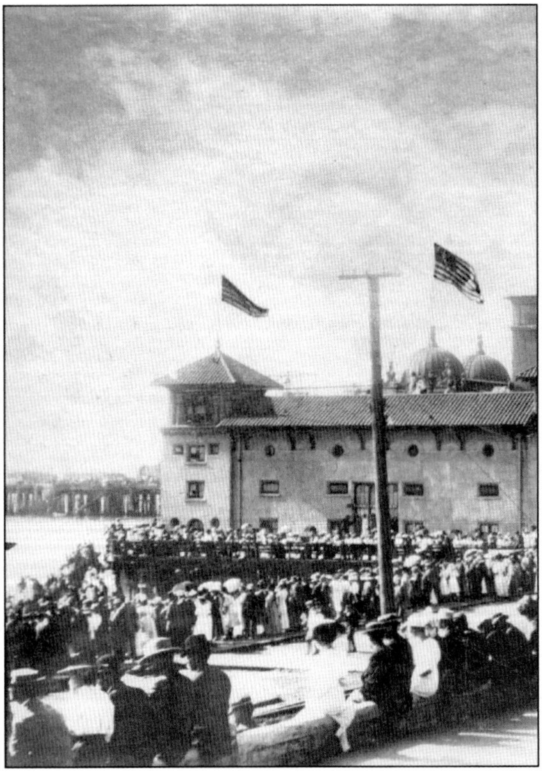

In June 1913, the *Los Angeles Times* reported that 20,000 people visited Redondo Beach for that year's official opening summer season. Both the Pacific Electric Red Cars and the Santa Fe Railway brought people from as far as San Bernardino and Redlands. The *Los Angeles Times* claimed that at least 2,000 automobiles lined Pacific Avenue and various side streets. Visitors enjoyed the Pavilion, the Lightning Coaster, band concerts, diving shows at the Plunge, and other activities, such as bowling, baseball, and relaxing on the beach.

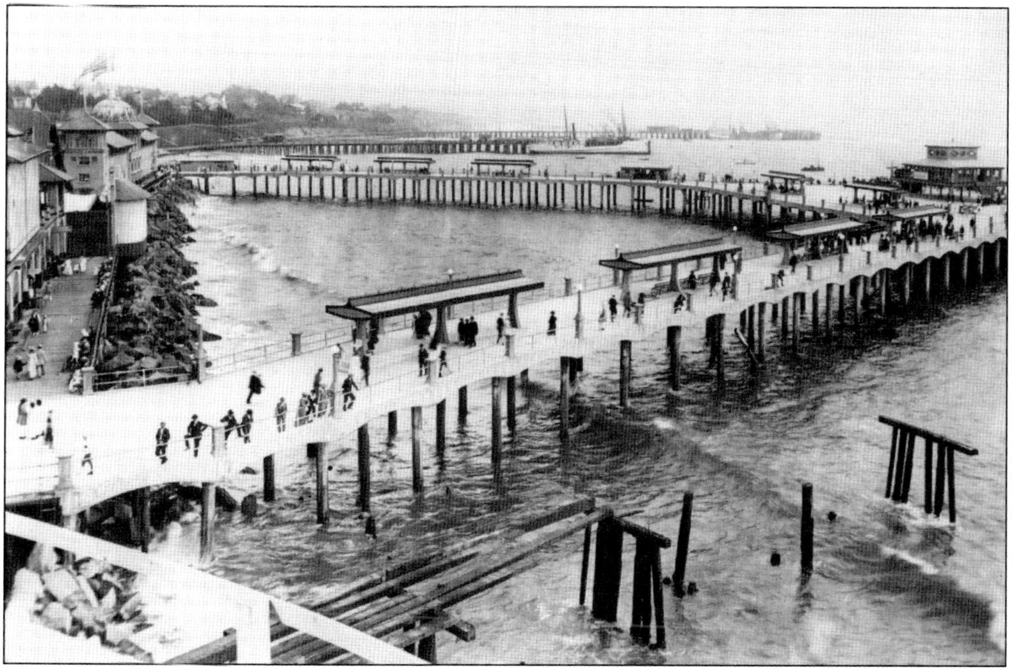

In April 1915, citizens voted to construct a "V"-shaped pleasure and fishing pier that extended 900 feet into the Pacific Ocean. The north entrance of the pier was to be placed just north of the Pavilion, and the south entrance was located in front of the Bath House. Other options for this vote included a straight pier at the foot of Diamond Street, a straight pier south of the Garland Building, and a straight pier at the foot of Emerald Street (where Wharf No. 1 once stood), as well as a horseshoe-shaped pier. Out of a possible 1,189, there were 744 votes for the triangular-shaped pier. Engineers reported that this first pier cost $120,000.

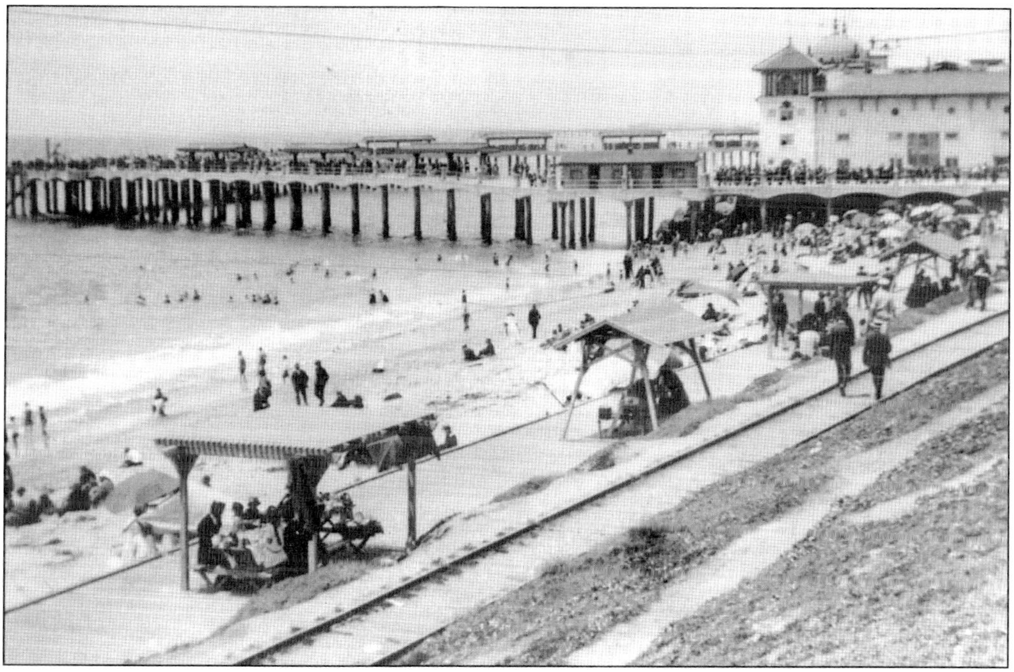

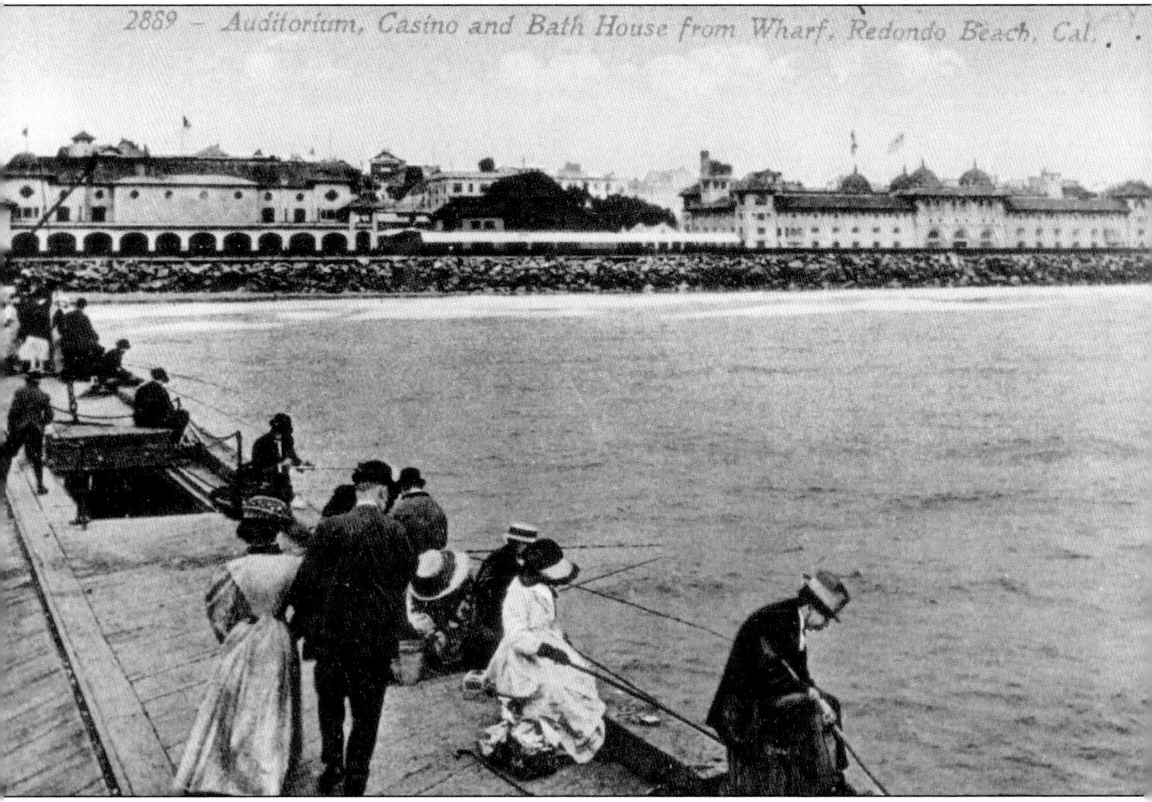

This postcard from the late 1800s shows the Plunge, casino building, and Pavilion from Wharf No. 1. As large entertainment venues typical to the pleasure pier era of the early 20th century, both of these buildings were prominently situated along the shoreline, capitalizing on the atmosphere of a fun fantasyland that was the beach.

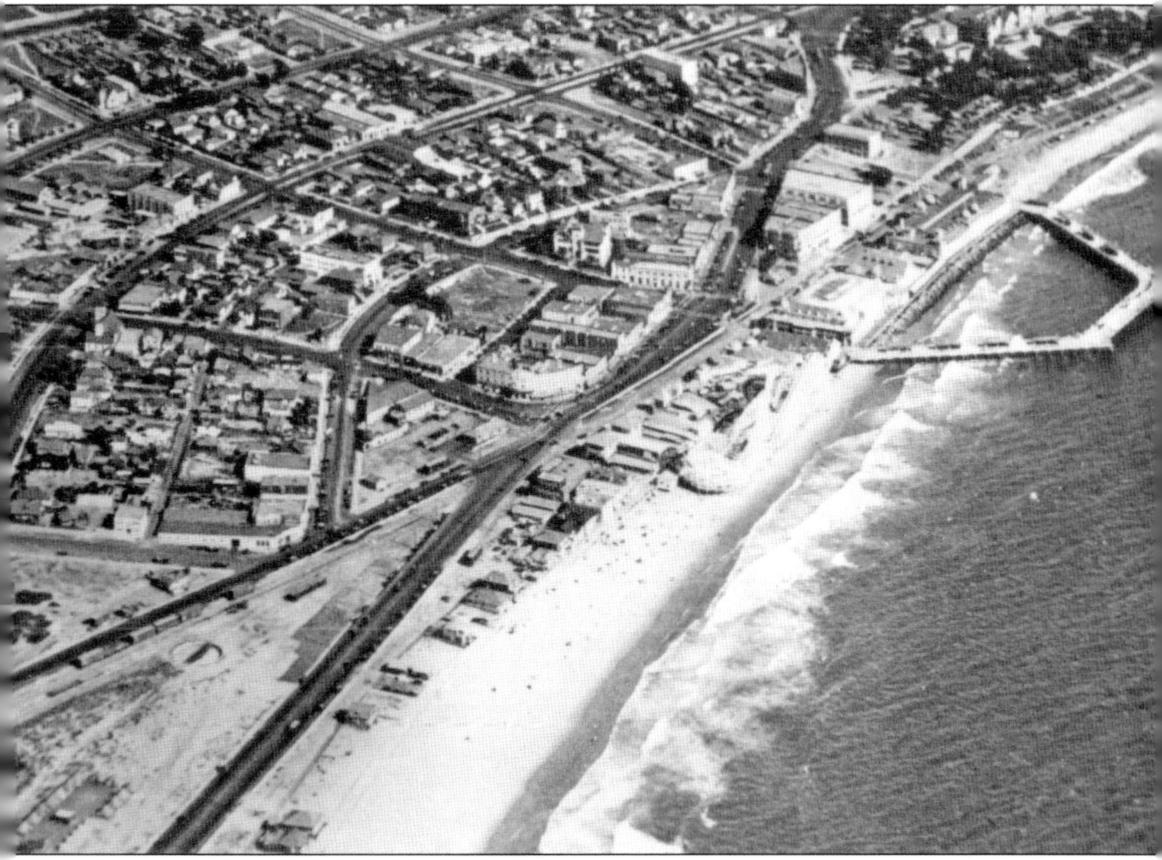

On April 30, 1915, a huge storm destroyed Wharf No. 1, which had been retired as a port due to its inadequacy for shipping needs. Articles from the *Los Angeles Times* explain that it was a popular pier for fishing in later years. After it was demolished, a new recreation pier was constructed. When the Endless Pier was built in 1916, it was the first of its kind in style and shape on the Southern California coast. Pictured above is the Endless Pier with the Giant Dipper roller coaster located to the north of it right on the shoreline.

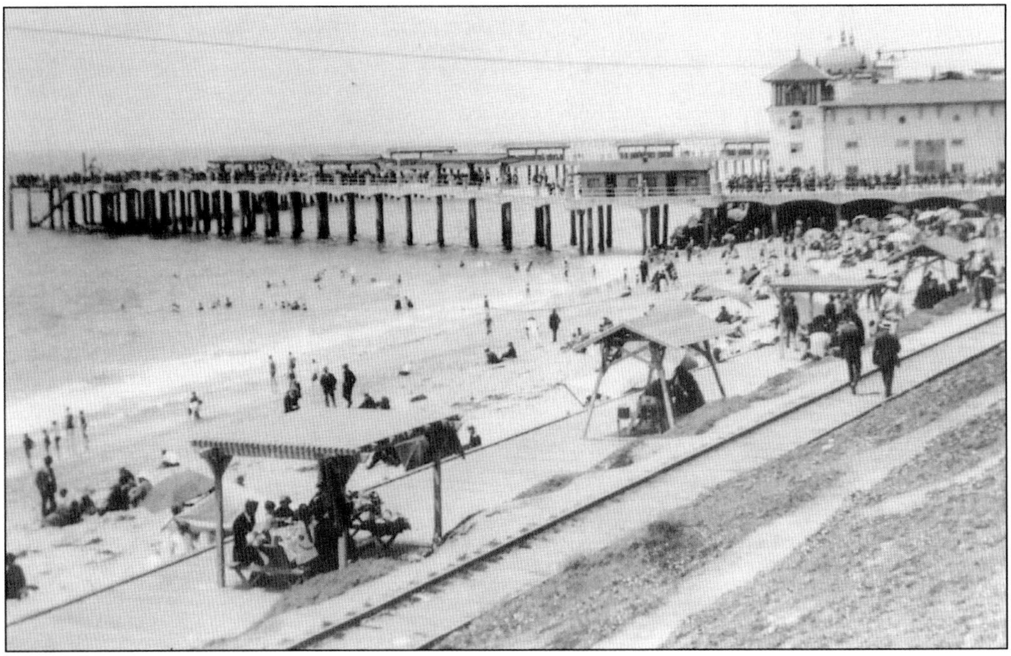

On August 5, 1916, more than 7,000 people attended the grand opening of the new municipal pier in Redondo Beach. At the ceremony were citizens from neighboring towns such as Torrance, Moneta, and Bridgedale. An orchestra played and a banquet was held. According to a *Los Angeles Times* article from 1916, two girls in "tight-fitting bathing suits" dove off the end of the pier and were forced to wear barrels as they exited the water. These two photographs show the new Redondo Beach pier. The image above was taken from the south side of the pier around 1916. Below is a postcard of promenaders along the pier around the same time period.

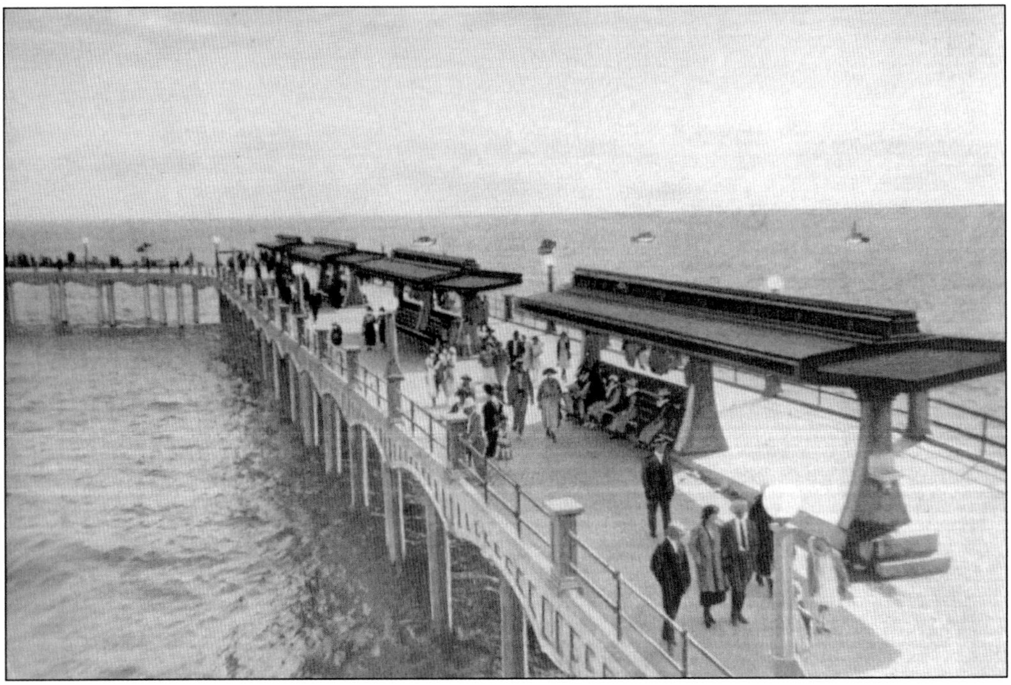

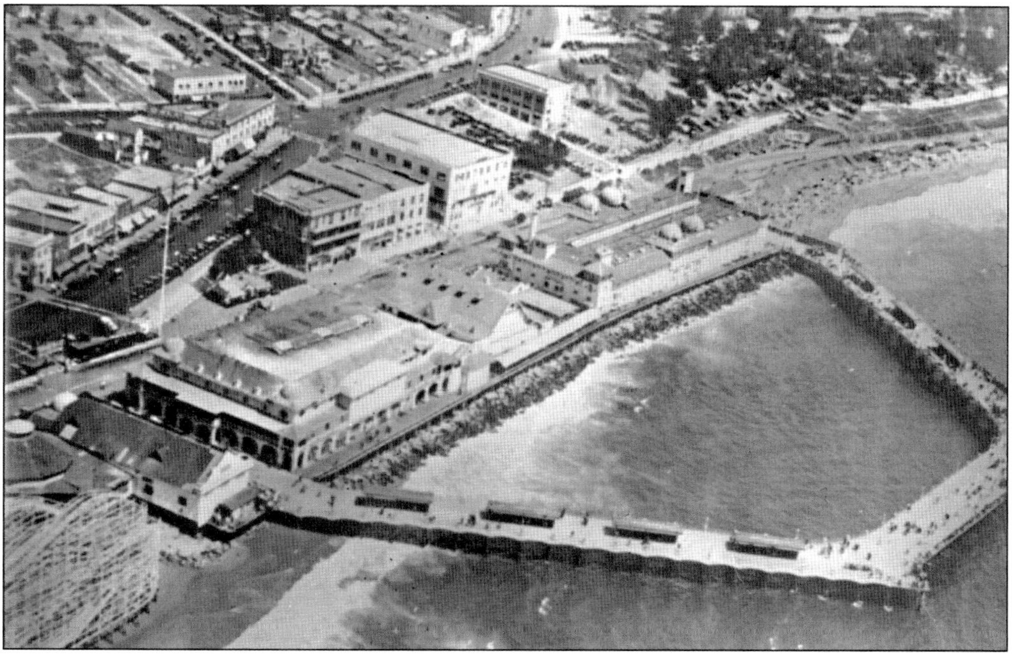

The new pier, called the Endless Pier, was constructed of reinforced concrete in the shape of a triangle. At the apex of the pier on the oceanside, a two-story sun parlor and observation pavilion was constructed, and at various points along the pier, shaded benches were placed for people to sit and look out over the water. There was also an outer fishing deck built along the sides of the pier so that promenaders were not disturbed with rogue fishing lines. However, in a matter of a few months, this Endless Pier was the victim of a storm that washed away some of its pilings. In October 1916, a new horseshoe-shaped pier was under construction, and it was ready by the spring of 1917. This time, the pier was reinforced with steel pilings that weighed 16 to 20 tons each, and the deepwater pilings weighed even more.

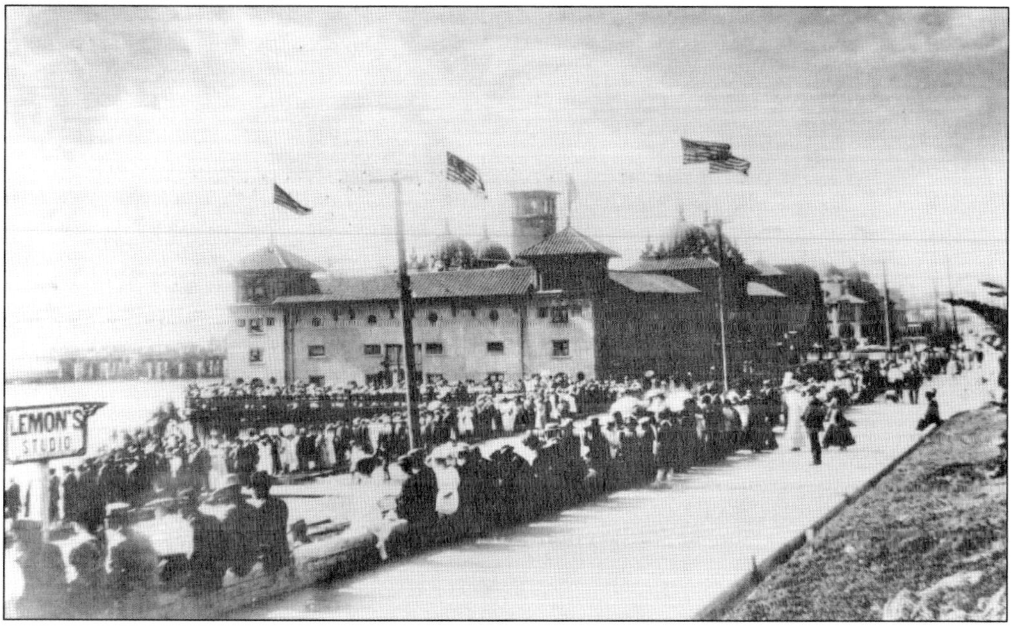

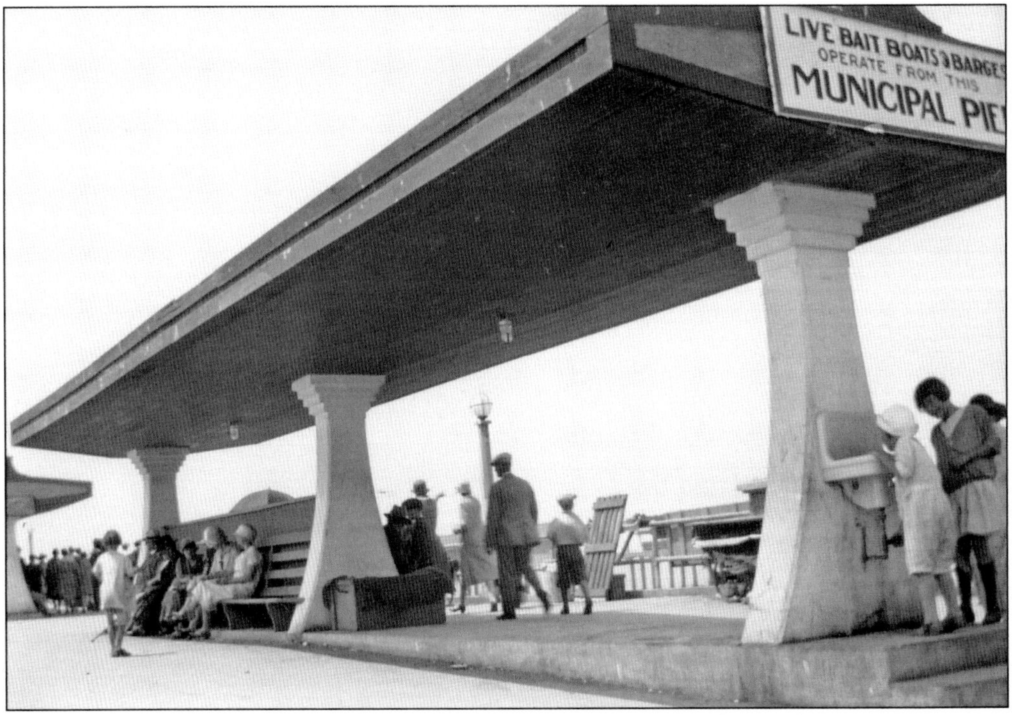

Pictured is a shaded settee with benches on the early pier in Redondo Beach.

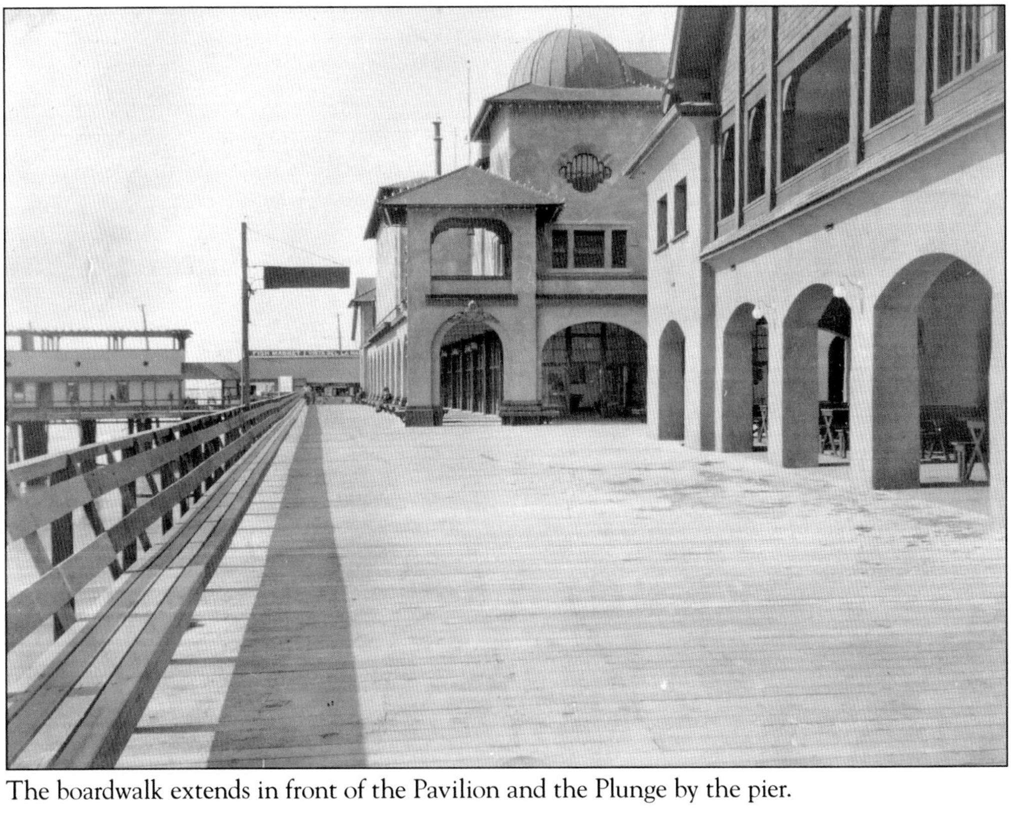

The boardwalk extends in front of the Pavilion and the Plunge by the pier.

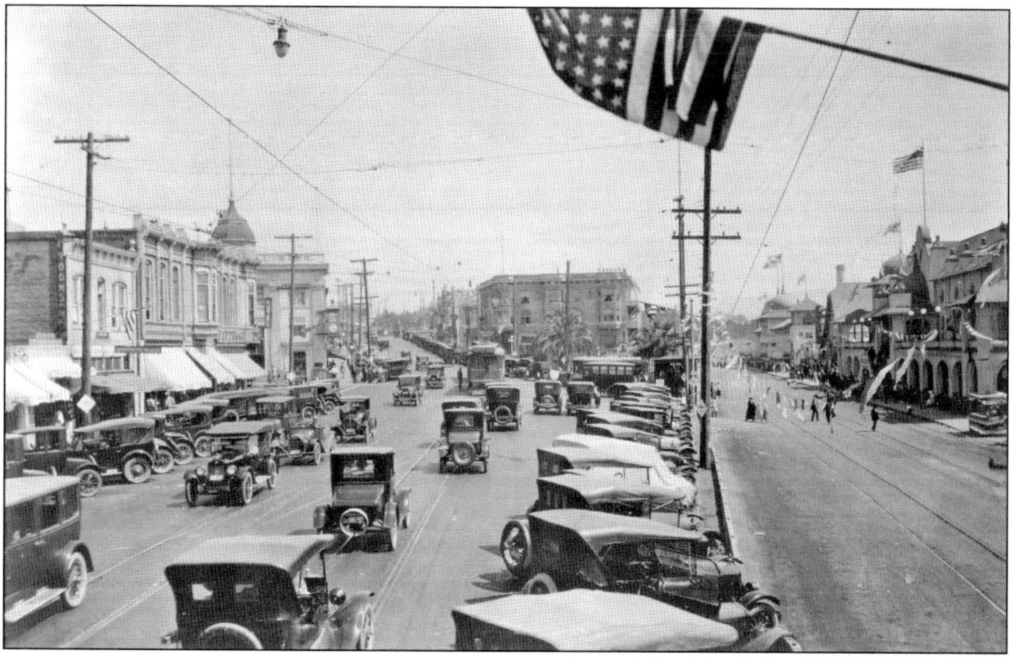

El Paseo became more popular with tourists after the addition of a new pleasure and fishing pier. Pictured above is El Paseo some time around 1915. The American flag is flown from most of the buildings on El Paseo, a sign of American patriotism in the early 20th century. Henry Huntington typically ceremoniously celebrated the arrival of his Pacific Electric Red Cars to new towns on the Fourth of July. Below is a beach scene in Redondo taken in the early 1900s.

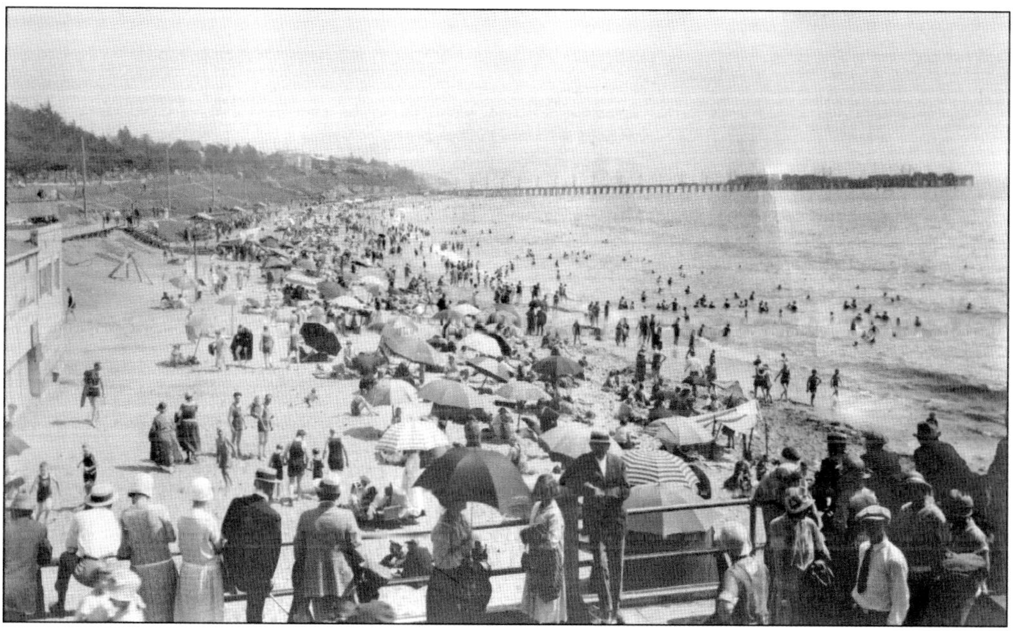

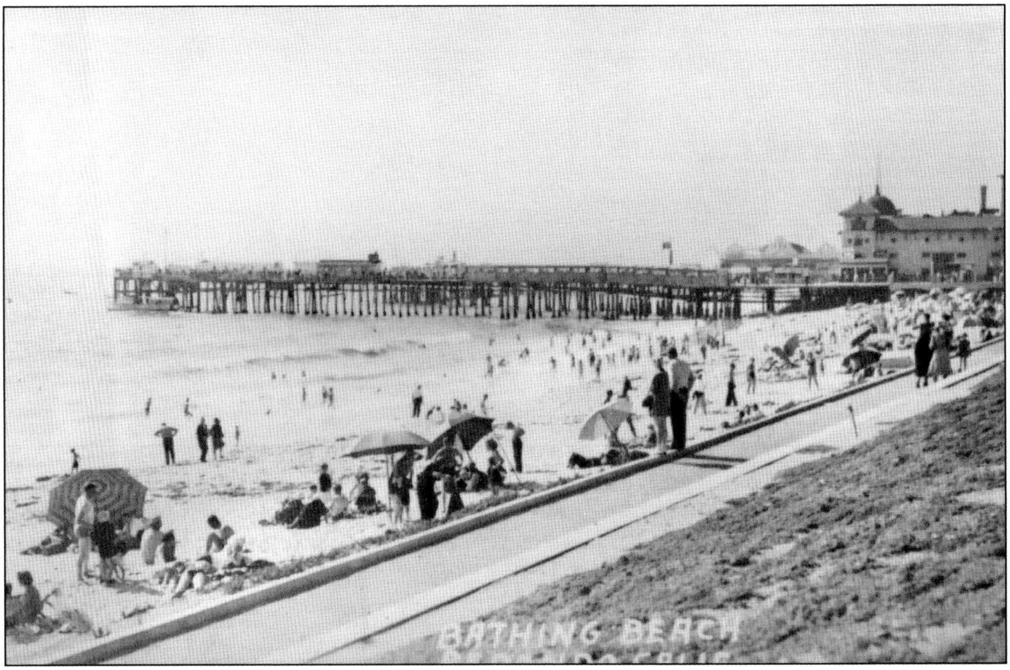

Groups of beachgoers enjoy a fair-weather day in the early 1900s. Seen in the distance are the Plunge and the Endless Pier.

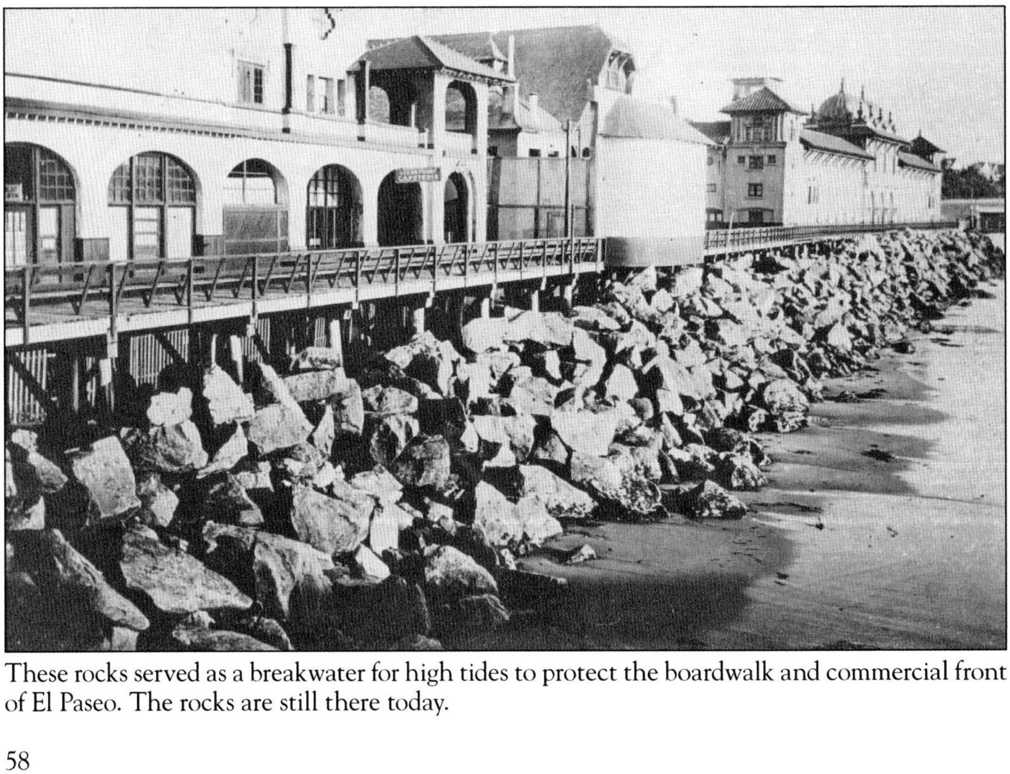

These rocks served as a breakwater for high tides to protect the boardwalk and commercial front of El Paseo. The rocks are still there today.

Shown at the right is a view of El Paseo from around 1920. The storefronts are very identifiable with their large hanging signage. The image below is of a candy kiosk on El Paseo in the 1920s. This is an example of the types of food vendors that adorned El Paseo. On the sign, it reads, "Peanuts, Pop Corn, Candies, Soda Pop." Redondo Beach had weekend-long pier and harbor celebrations in the summer. According to a *Los Angeles Times* article, festivities included "fancy diving" off the pier and in the Bath House, concerts by the Redondo Band, vaudeville performances, a balloon shower "in which many thousands of balloons were released," and "fancy dancing" in the auditorium of the Pavilion. In August 1917, as many as 30,000 people reportedly attended pier and harbor celebrations in Redondo.

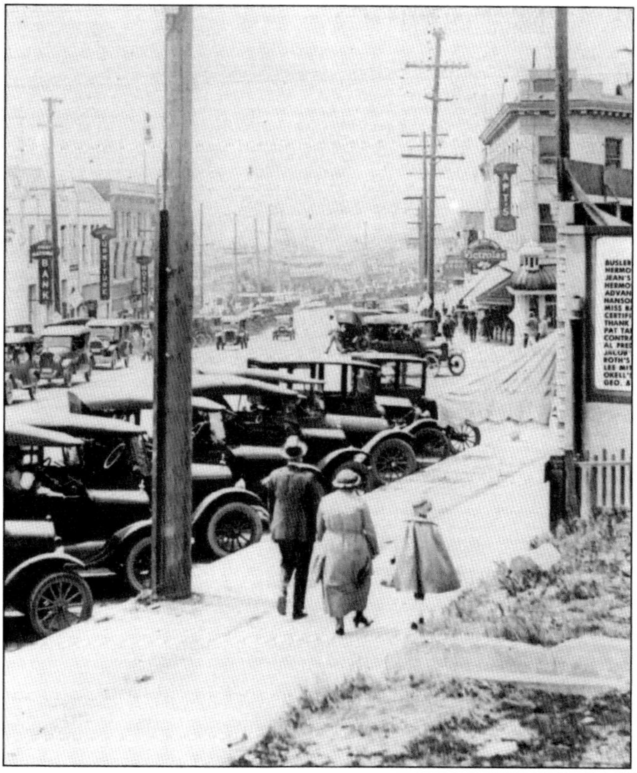

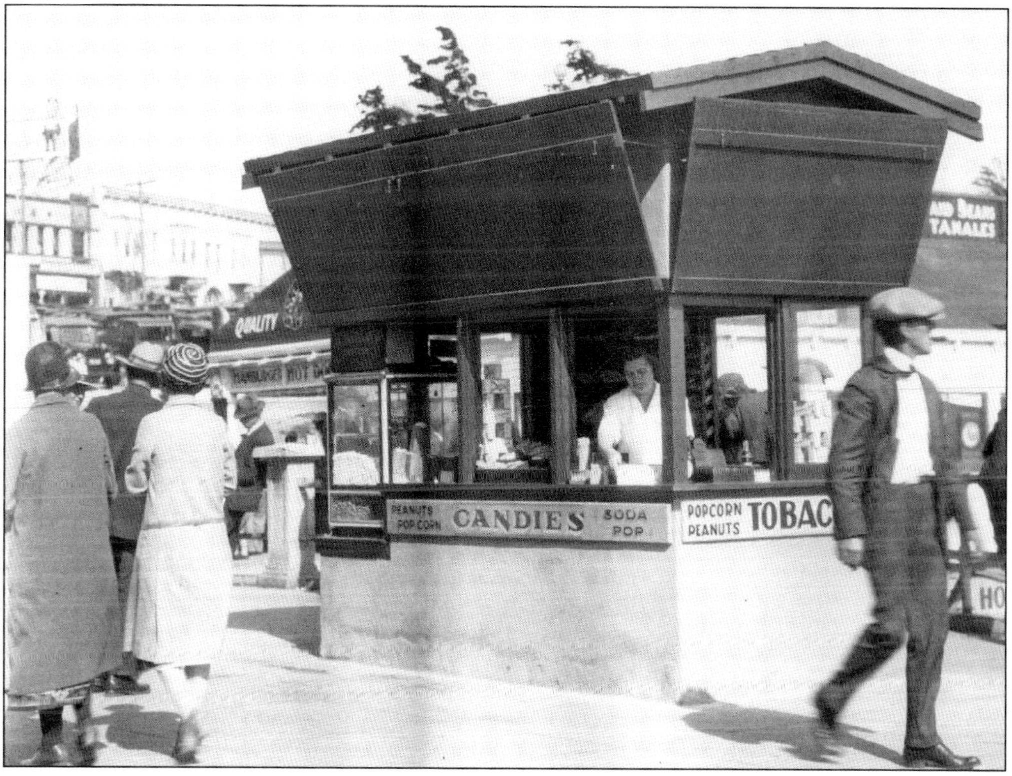

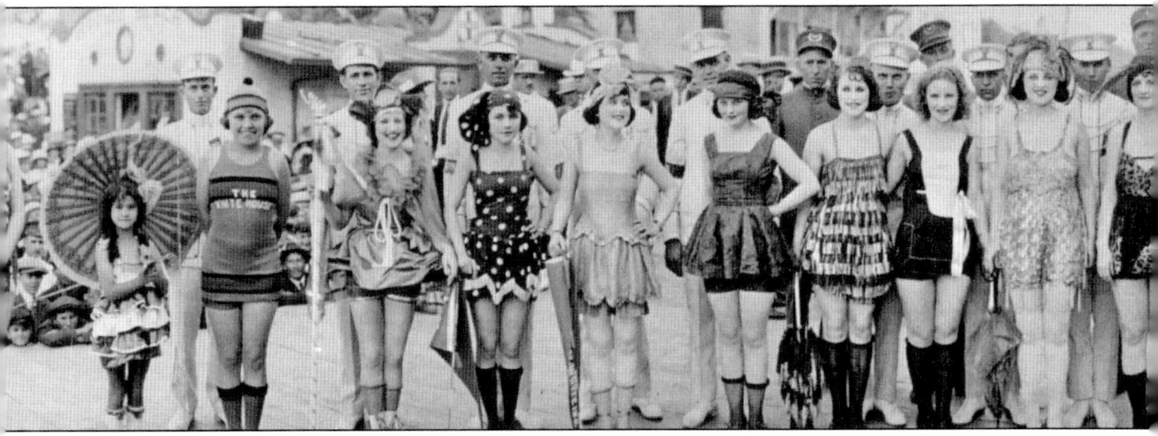

One of the most popular events at the beach was the bathing beauty contest, which began in the early 1900s. However, coming out of the Victorian era of clothing, when it was proper to wear full-length dresses and long sleeves to the beach, not all were enthusiastic of this newfangled wardrobe. In Long Beach in 1920, a city official ordained that adult swimwear must adhere to this rule: "Completely conceal from view each leg from the hip joint to a line around the leg one-third of the way to the knee." Although this regulation was repealed six years later, the city held a bathing beauty parade, and a local minister called the city council "vampires of human flesh." This is a beauty contest in which the swimwear looked more like costumes than actual bathing suits. This photograph was taken around 1925 on El Paseo.

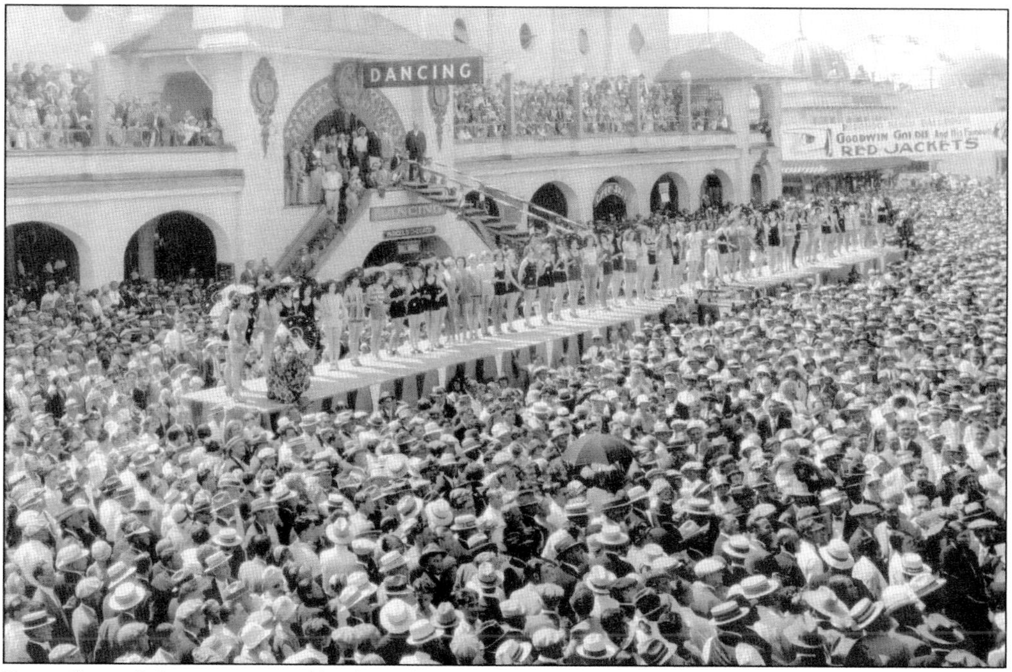

Regardless of the criticism of bathing beauty contests, they remained popular. These pageants were later noticed by automobile dealers, who then hired bathing beauty contestants to show off new car models in various festivals and parades held in cities all over Southern California. This is a 1927 photograph of a beauty contest in front of the Redondo Pavilion. The crowd consists mostly of men.

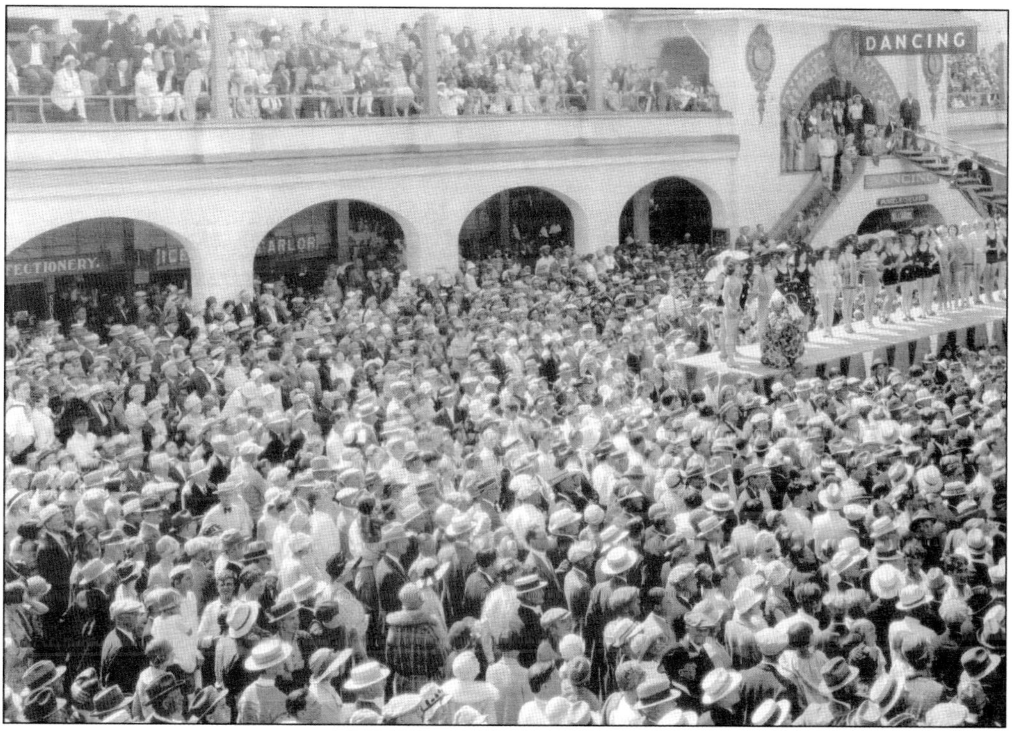

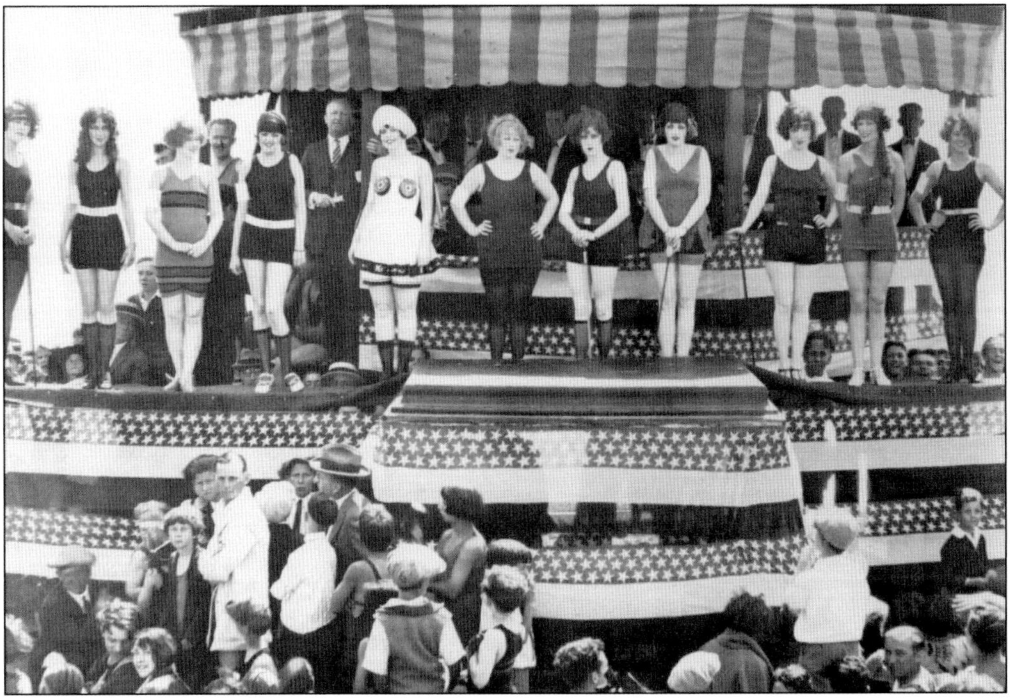

Contestants in these pageants dressed up in costume-like attire that was not suitable for swimming in the hopes of winning cash prizes. According to a *Los Angeles Times* article, winners fulfilled the criteria of being "the most shapely and beautiful" of all competitors. Often, famous motion picture stars and artists were the judges. The pageants typically entailed parades as well. Aspiring actresses and models participated in these well-known beachfront events.

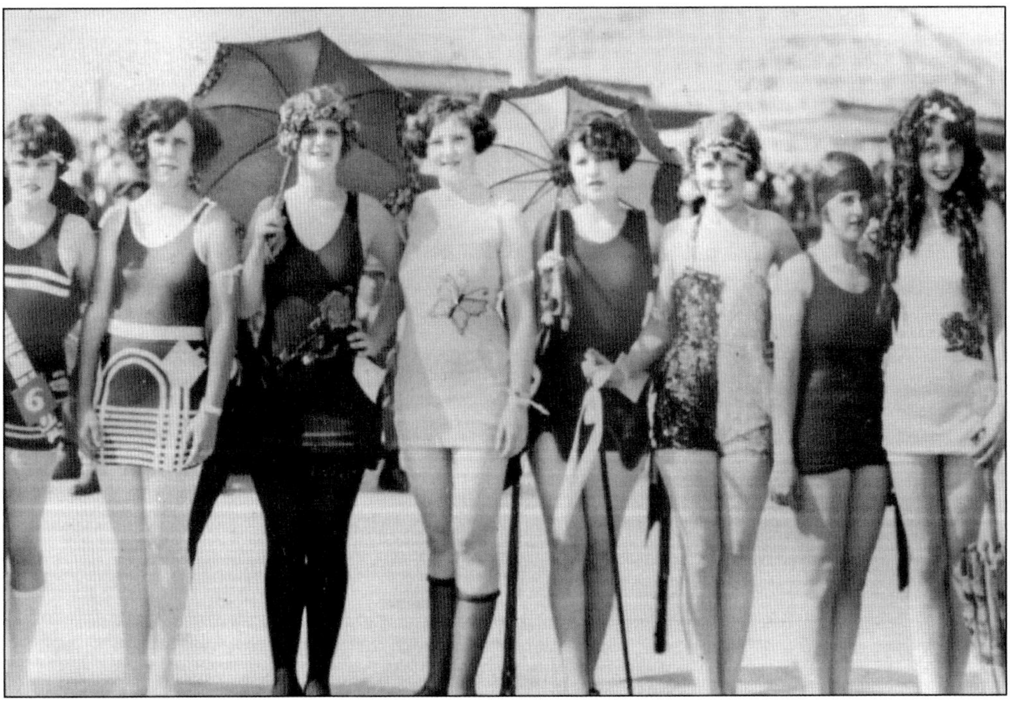

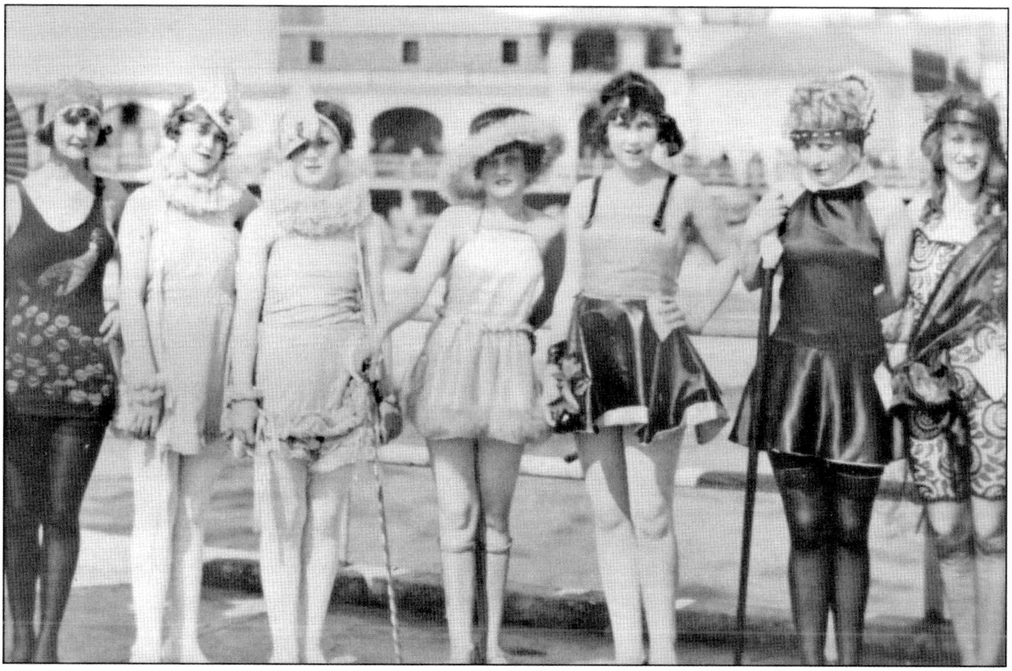

In these 1924 photographs are a bathing beauty pageant, seen above, and a costume contest, seen below, at Redondo Beach. In the picture below, the second girl from the left was the winner.

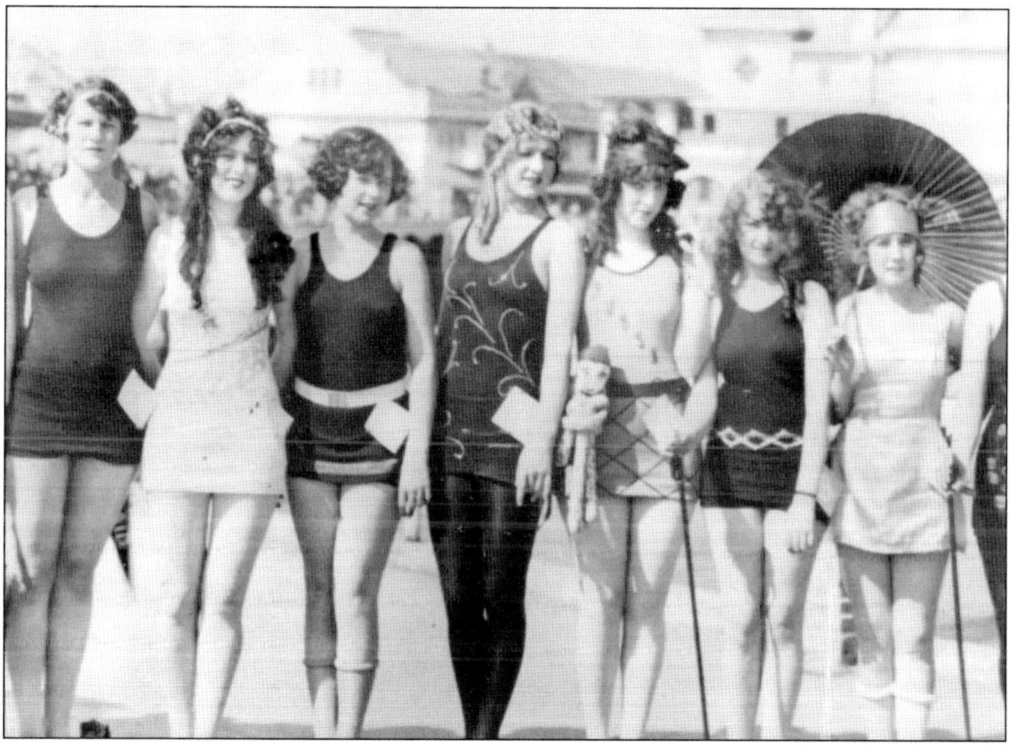

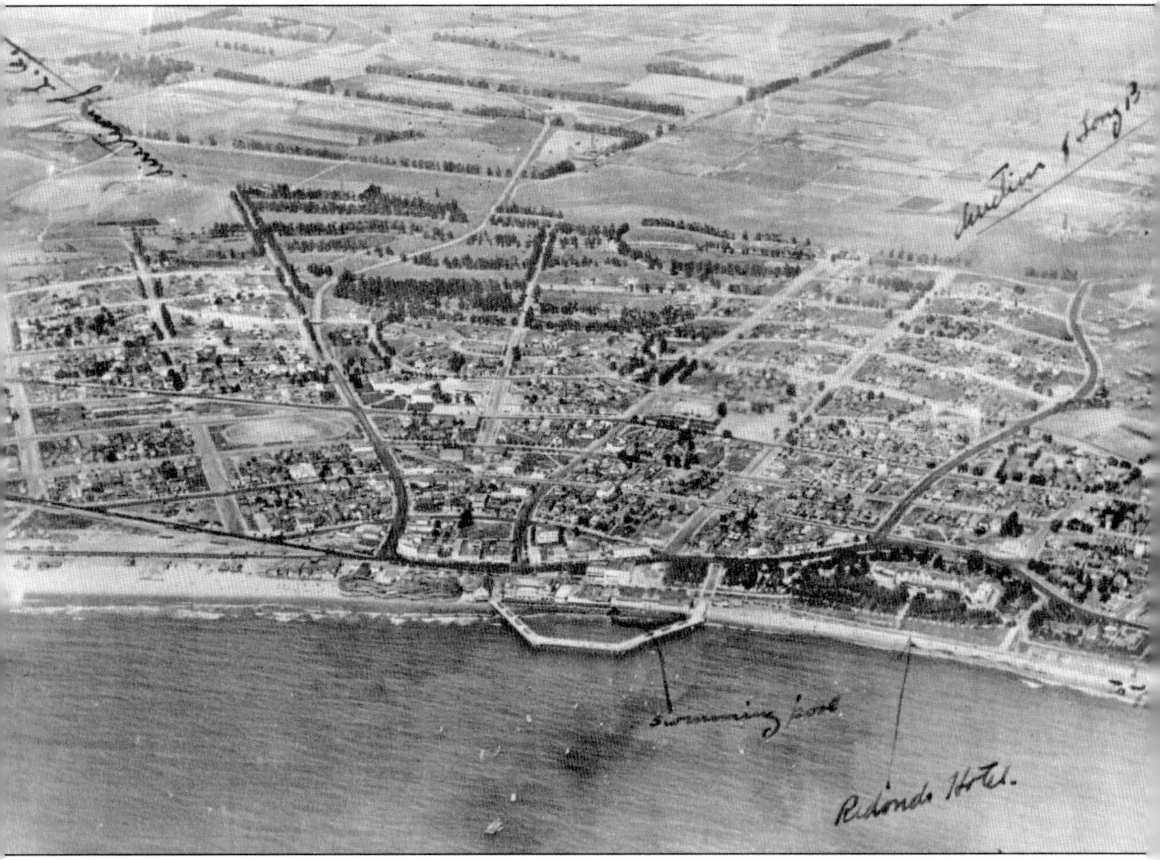

This is an aerial photograph of Redondo Beach taken some time around 1917. Horseshoe Pier has been constructed, and wharves No. 1 and No. 2 are gone. Wharf No. 1 would have been at the north end of Horseshoe Pier, and Wharf No. 2 would have been somewhere in front of Hotel Redondo, which is still extant in this picture. El Paseo, the Pavilion, the Plunge, and the Lightning Racer are all discernible, as well as open farmland in the distance. The town's original layout is based on the Chautauqua movement of the late 19th century. Established as an adult education movement in the United States, the Chautauqua's general philosophy was to bring entertainment and culture to the whole community. This was typically done with speakers, teachers, musicians, entertainers, preachers, and specialists of the day moving in circuits to locations all around the country. Redondo Beach had laid the original city plan in the shape of an oil lamp, representing a symbol of the Chautauqua philosophy, but that layout can barely be recognized today. The city also built a Chautauqua Meeting House in the shape of a decagon where the traveling circuits of speakers taught the masses.

This postcard shows two bathers enjoying the water at Redondo Beach. Behind them is a rope that was tied to large posts embedded deeply into the sand. The rope was designed to hold the weight of beach patrons who wanted to venture into the ocean but did not know how to swim.

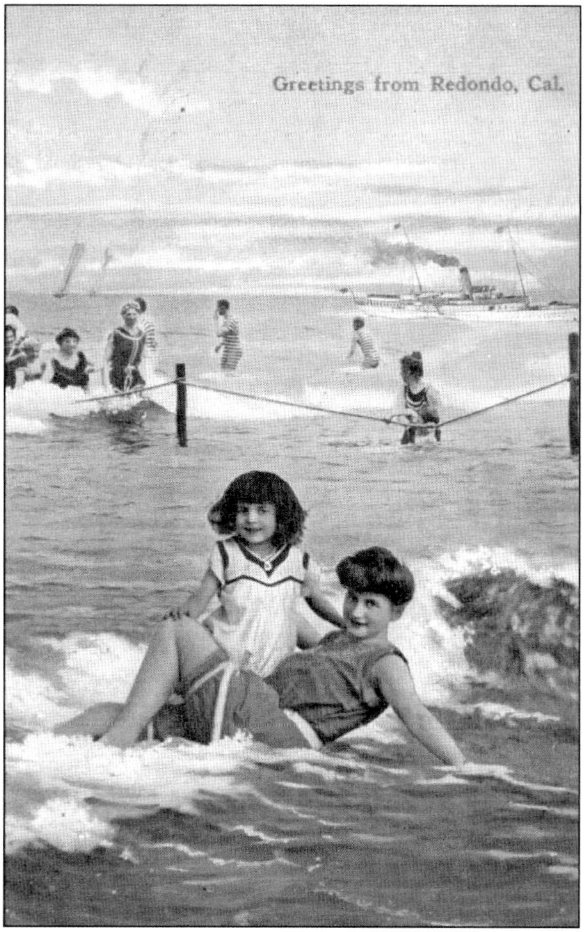

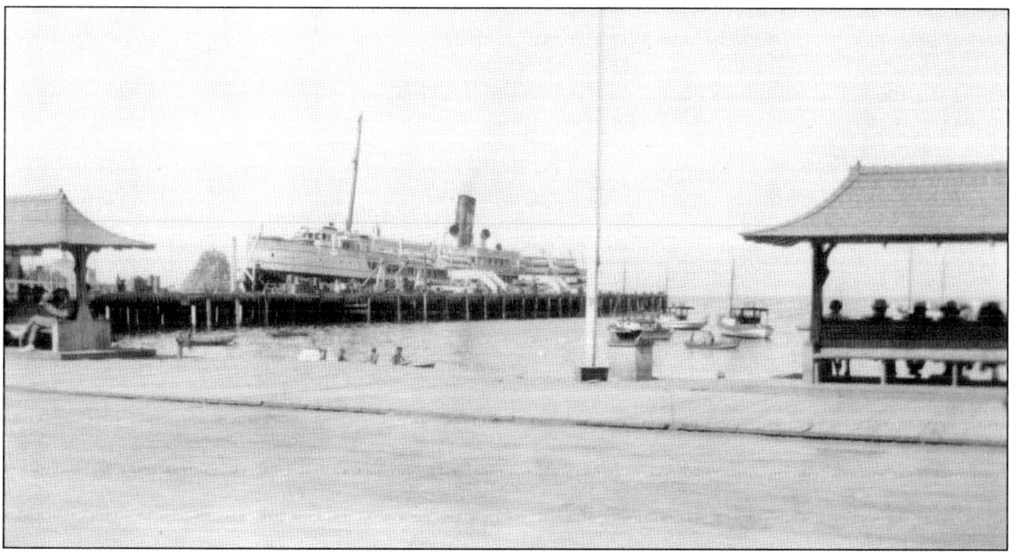

Pictured here is one of Redondo Beach's wharves in the early 1900s. This photograph was likely taken from El Paseo, which today is a marina and condominiums.

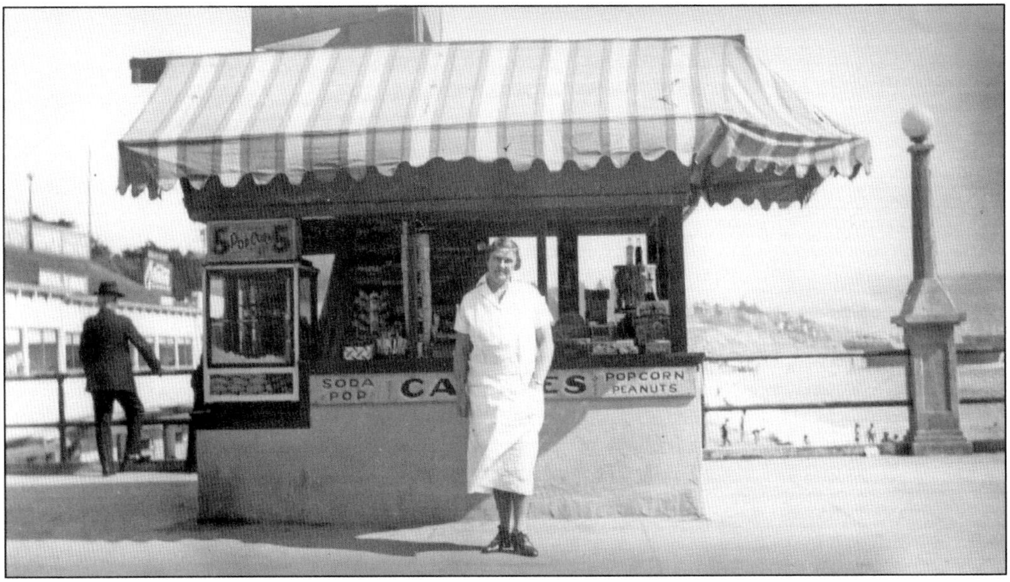

Some of the influences of the early beach resort era in Southern California were the amusement parks on the East Coast, as well as state and world fairs. Another influence was the concept of a midway, which was a main thoroughfare of rides, amusements, and concessions. Hence, the midway in Redondo Beach was located roughly between the Plunge and the Pavilion on El Paseo. Another influence was food. The first hot dog was sold at Coney Island. Deep-fried finger foods and snacks on a stick originated at state fairs and the 19th-century World's Fair expositions in the United States. In the early 20th century, concessions were important staples at Redondo Beach. Vendors sold hot dogs, corn dogs, popcorn, and cotton candy, as well as other greasy, fried, and sugary treats along El Paseo.

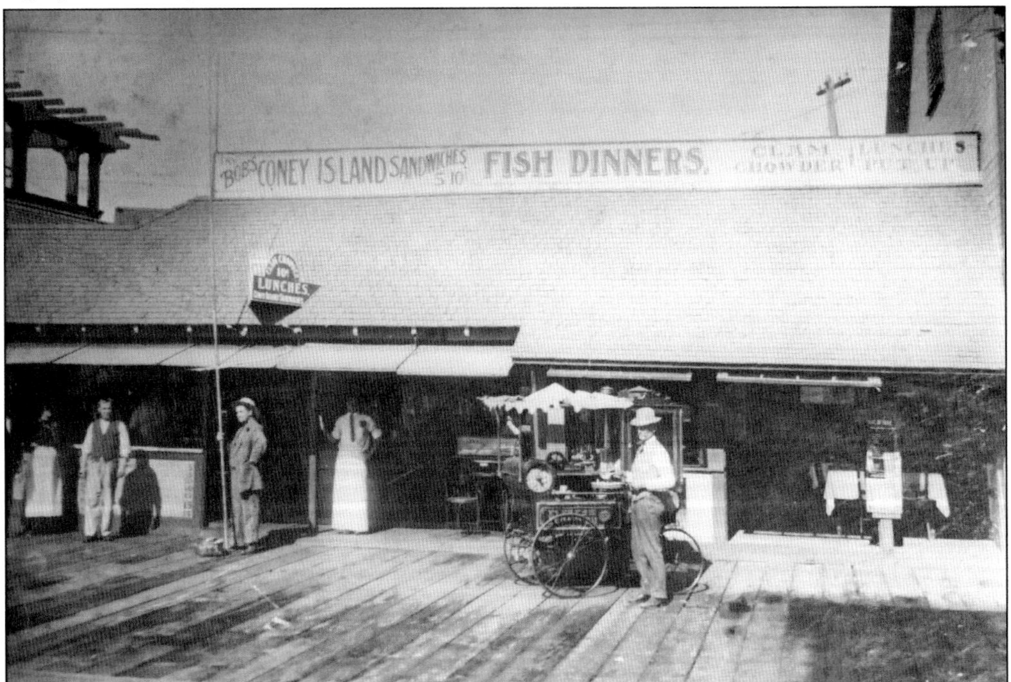

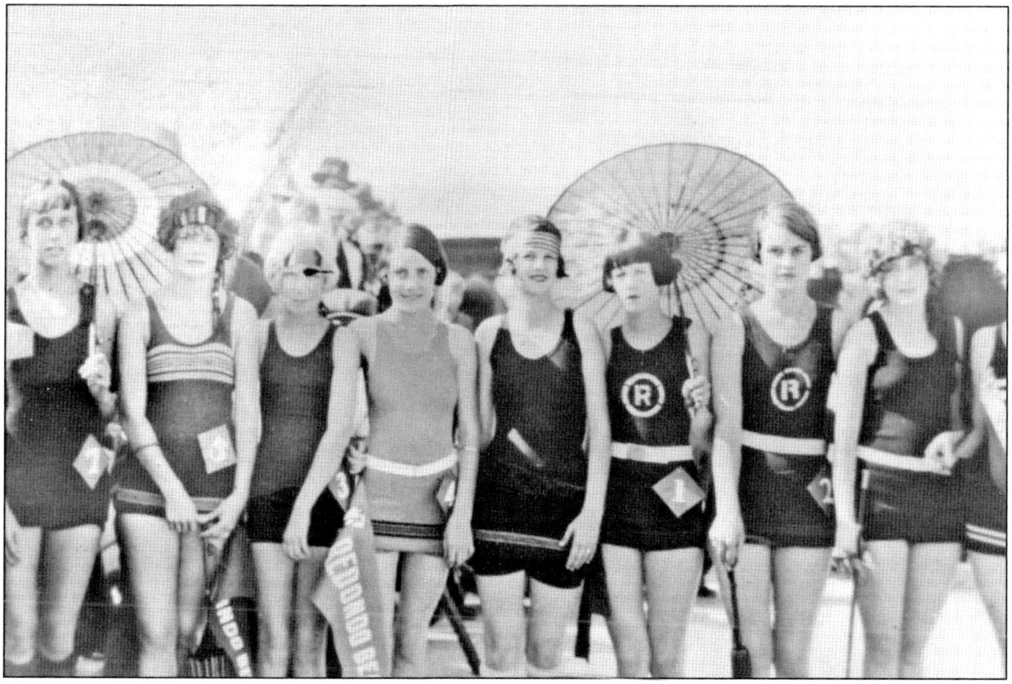

This is a photograph of either a beauty pageant or a swimming competition. The women in the bathing suits with the "R" have rented their suits from the Plunge.

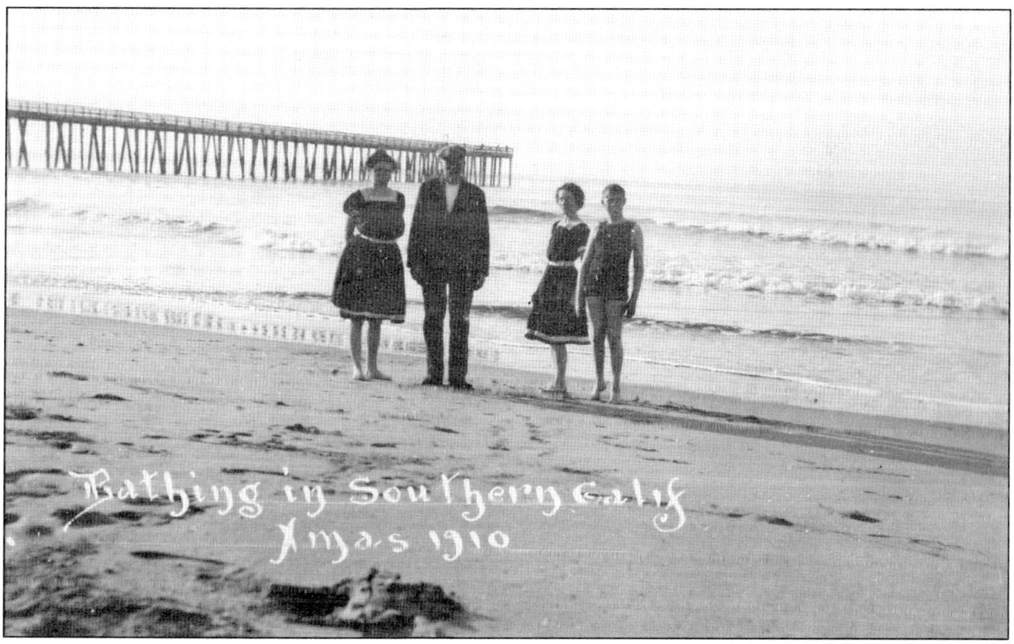

Above is a family photograph in which everyone except for the father is dressed in late-19th-century swimwear. Note how the women's bathing suits are knee-length full skirts.

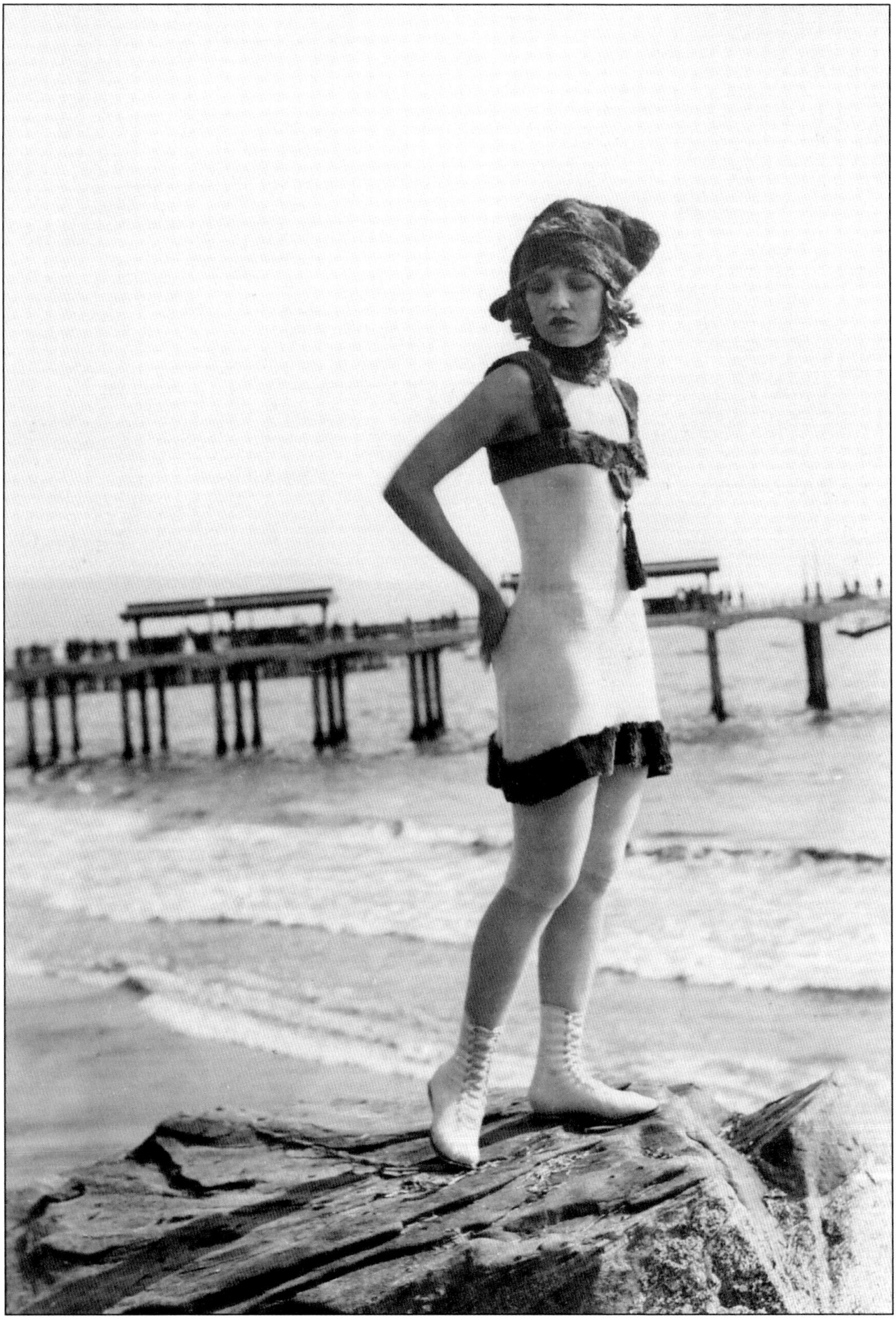

In this photograph is an aspiring model or beach beauty contestant from around 1916. In the background is the pier.

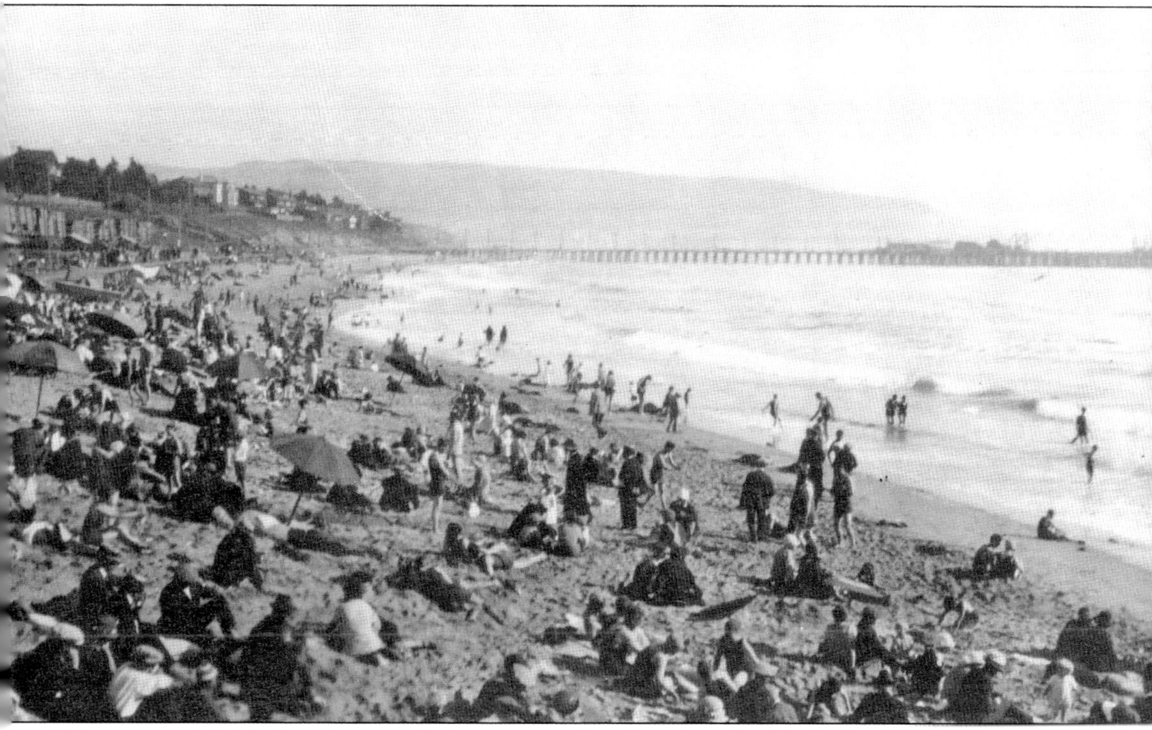

Taken around 1915, this photograph shows a crowd gathered at the beach. Seen in the distance is one of the wharves, and beyond that is the Palos Verdes Peninsula.

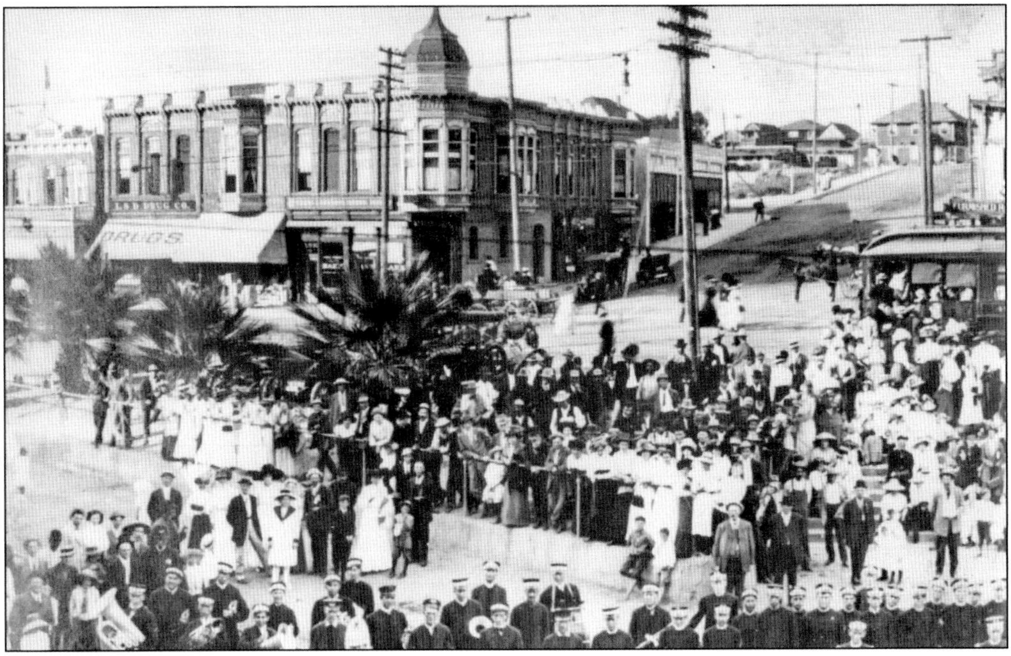

Because El Paseo was the main thoroughfare to the pier, Pavilion, and Plunge, occasional ceremonies were held nearby on Pacific Avenue and El Paseo. Pictured above is a ceremony at the intersection of Emerald Street and Pacific Avenue. There is a full band in uniform for the celebration. Below is a similar photograph taken south of the meeting point of Emerald Street and Pacific Avenue, specifically where Pacific Avenue and El Paseo split. The Santa Fe Railway train depot is just to the right, outside of the frame. These photographs were taken around 1925.

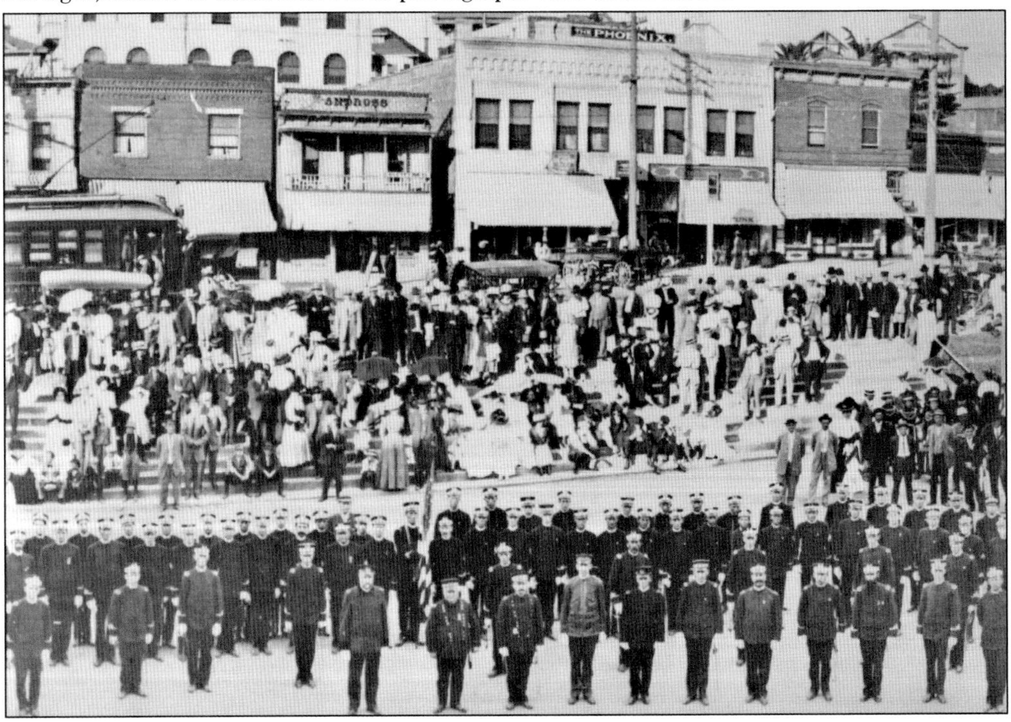

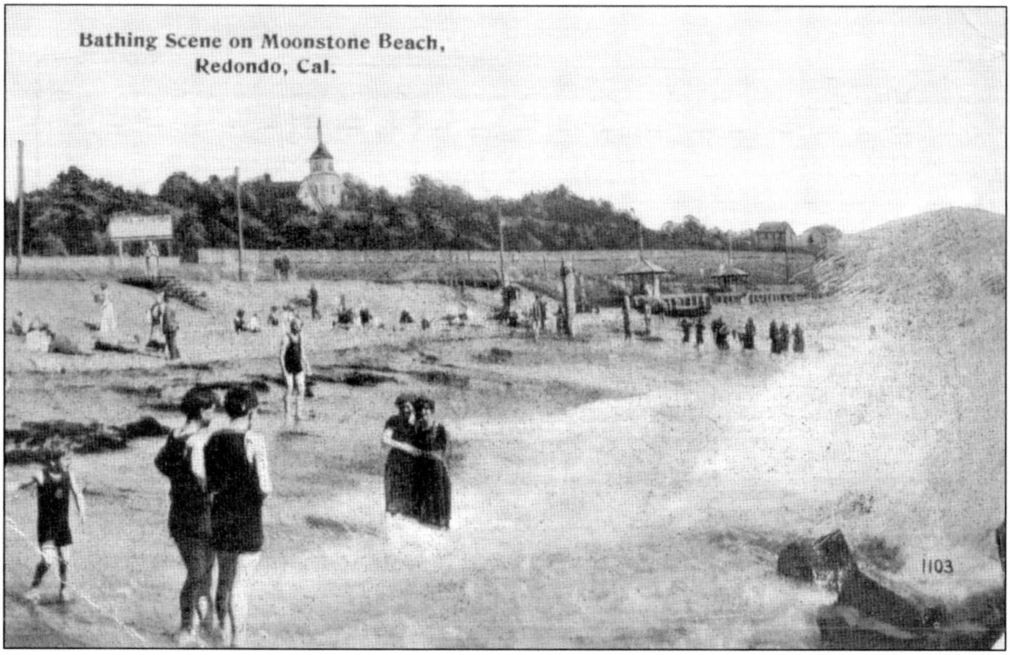

This is a postcard of some bathers enjoying the beach in the early 1900s at Moonstone Beach, located north of the Endless Pier and El Paseo amusement zone. (Courtesy of Donald Krintz.)

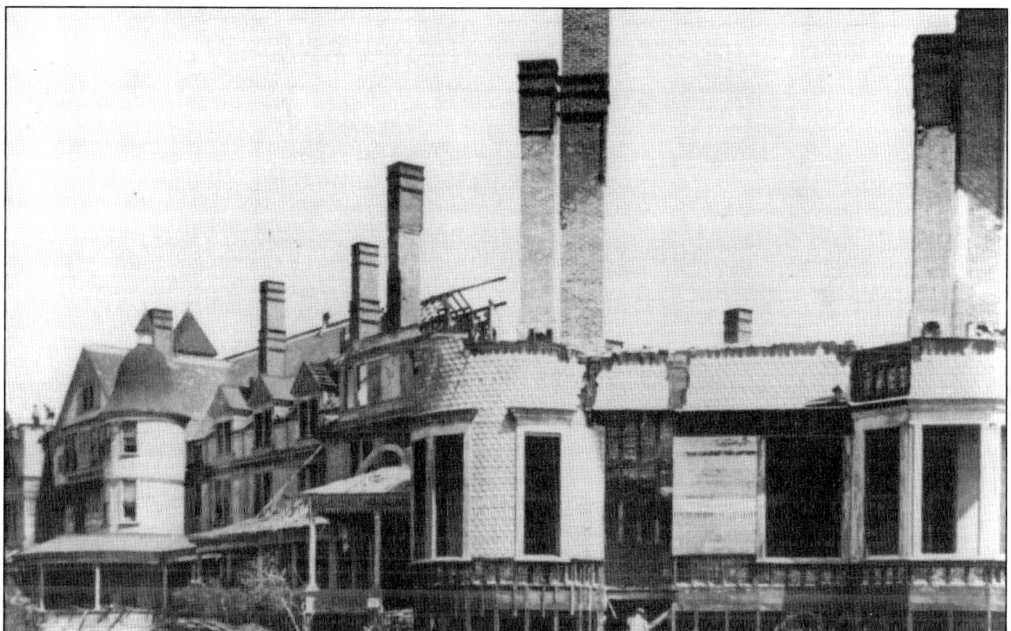

Unfortunately, the first Endless Pier was washed away in a storm, but it was rebuilt with wood pilings. The pier has been called several names, such as Endless Pier, Horseshoe Pier, and today, Municipal Pier. In 1925, Hotel Redondo was torn down, but this did not stop the tourism industry. Redondo's resident population was growing, as was the construction of new houses. In 1930, a library was built on the site of the former Hotel Redondo. It remains there today as the veterans center.

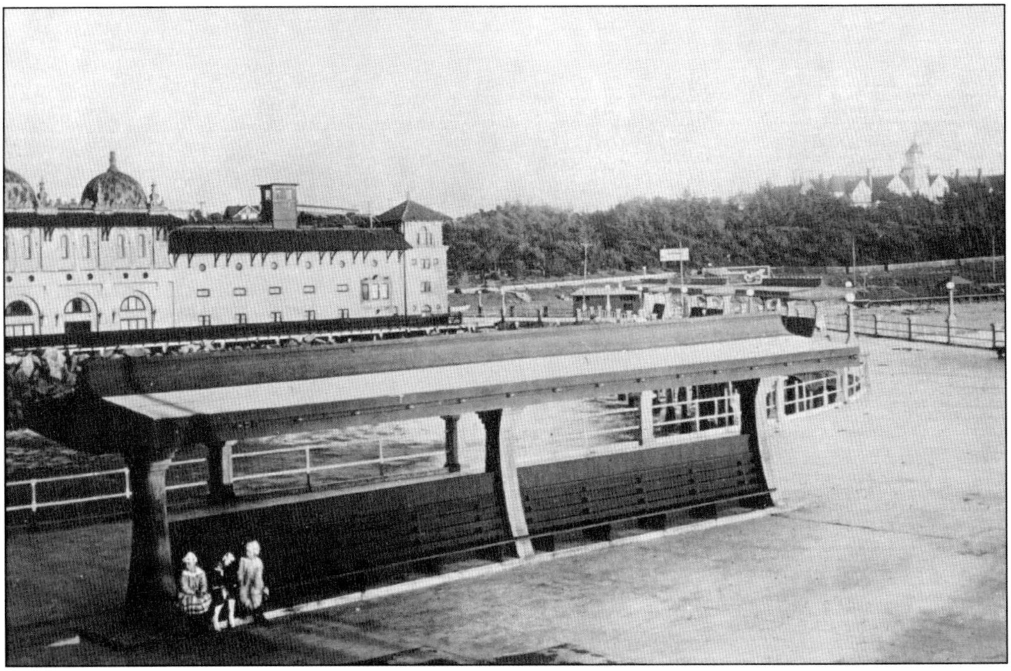

Children sit under one of the shaded settees on the pier. In the distance, Hotel Redondo is on the bluff, and the Plunge can be seen at the south end of the pier.

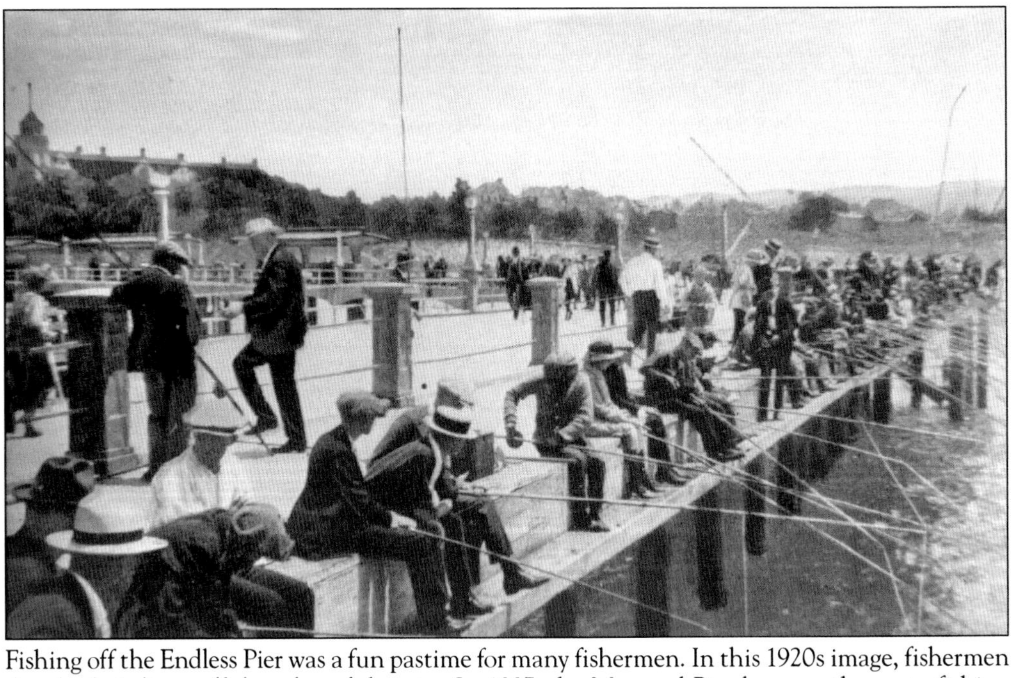

Fishing off the Endless Pier was a fun pastime for many fishermen. In this 1920s image, fishermen dangle their lines off the edge of the pier. In 1927, the Monstad Pier became the main fishing pier in Redondo.

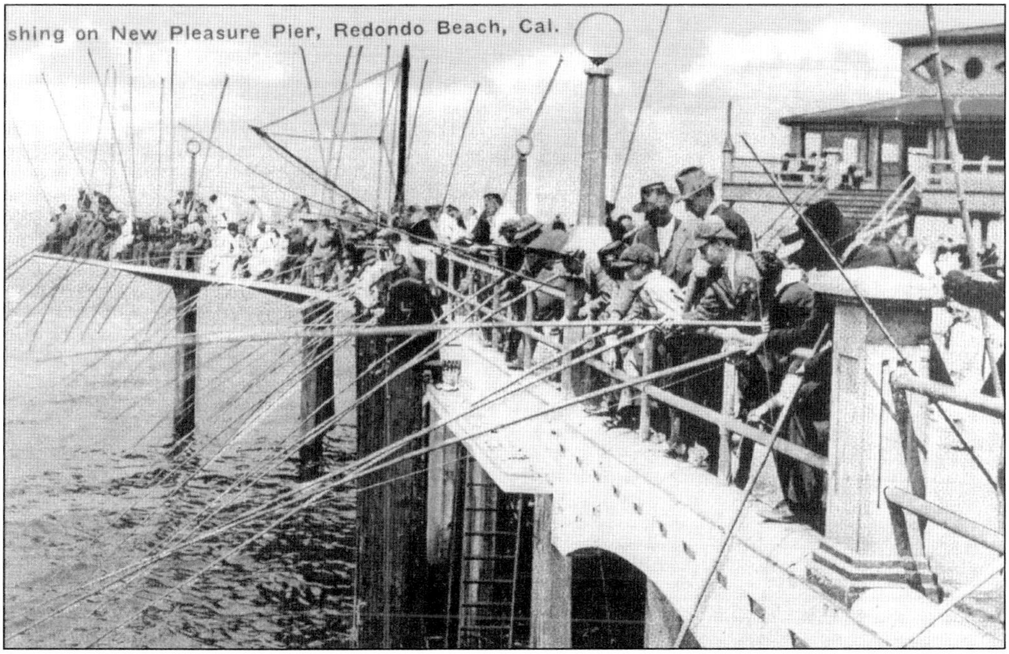

In 1925, the city approved Capt. Webb Monstad's request for the construction of a fishing pier adjacent to Horseshoe Pier. Monstad was a former merchant marine shipper. The pier was built from the south entrance of Horseshoe Pier and jutted 300 feet into the ocean. In 1936, the pier was extended farther by 100 feet.

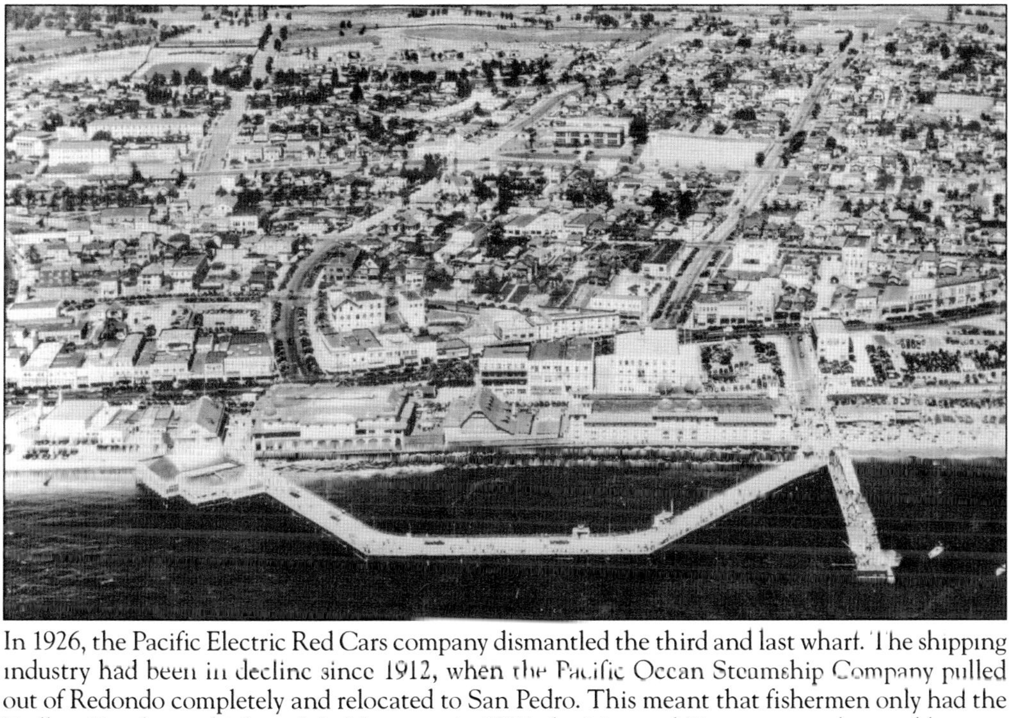

In 1926, the Pacific Electric Red Cars company dismantled the third and last wharf. The shipping industry had been in decline since 1912, when the Pacific Ocean Steamship Company pulled out of Redondo completely and relocated to San Pedro. This meant that fishermen only had the Endless Pier from which to fish. However, in 1927, the Monstad Pier was complete and became the main fishing pier in Redondo Beach.

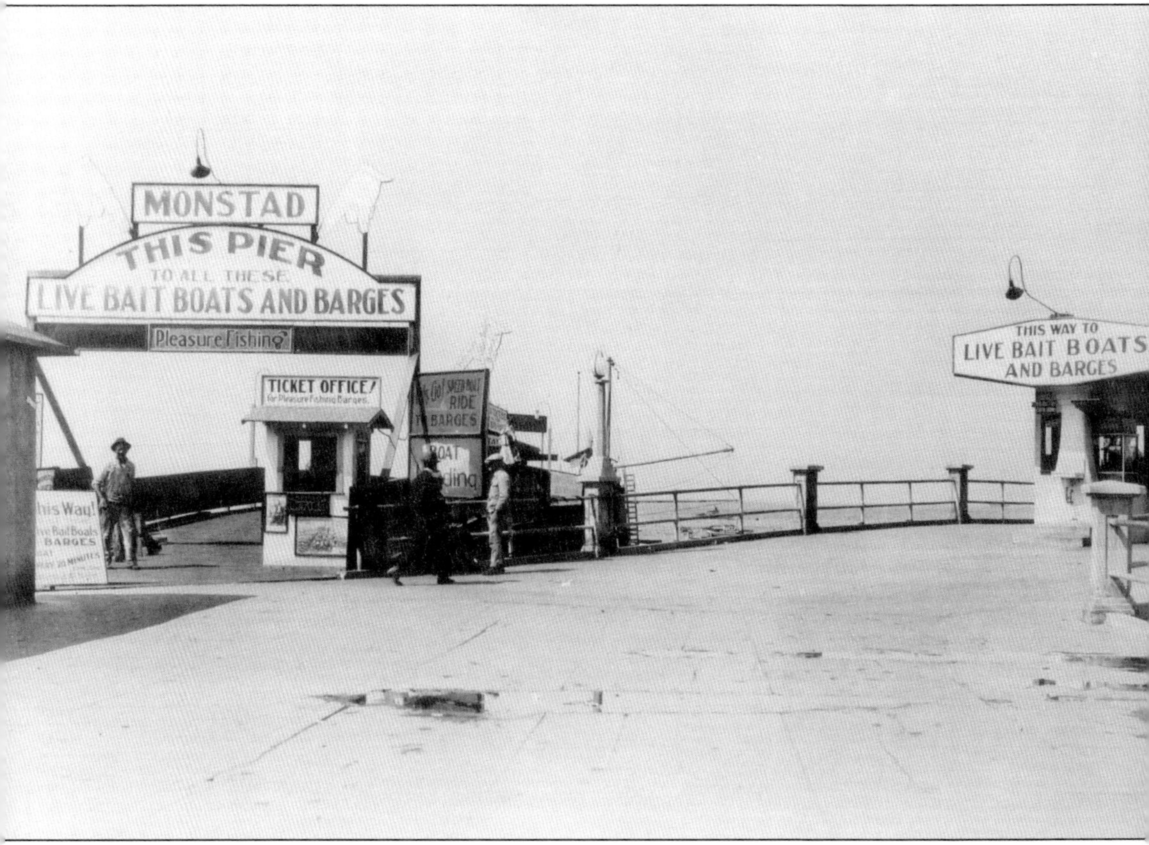

The sign at the entrance to the Monstad Pier reads, "Monstad This Pier To All These Live Bait Boats and Barges." The sign has two hands pointing down at the lettering. This photograph was taken around 1930.

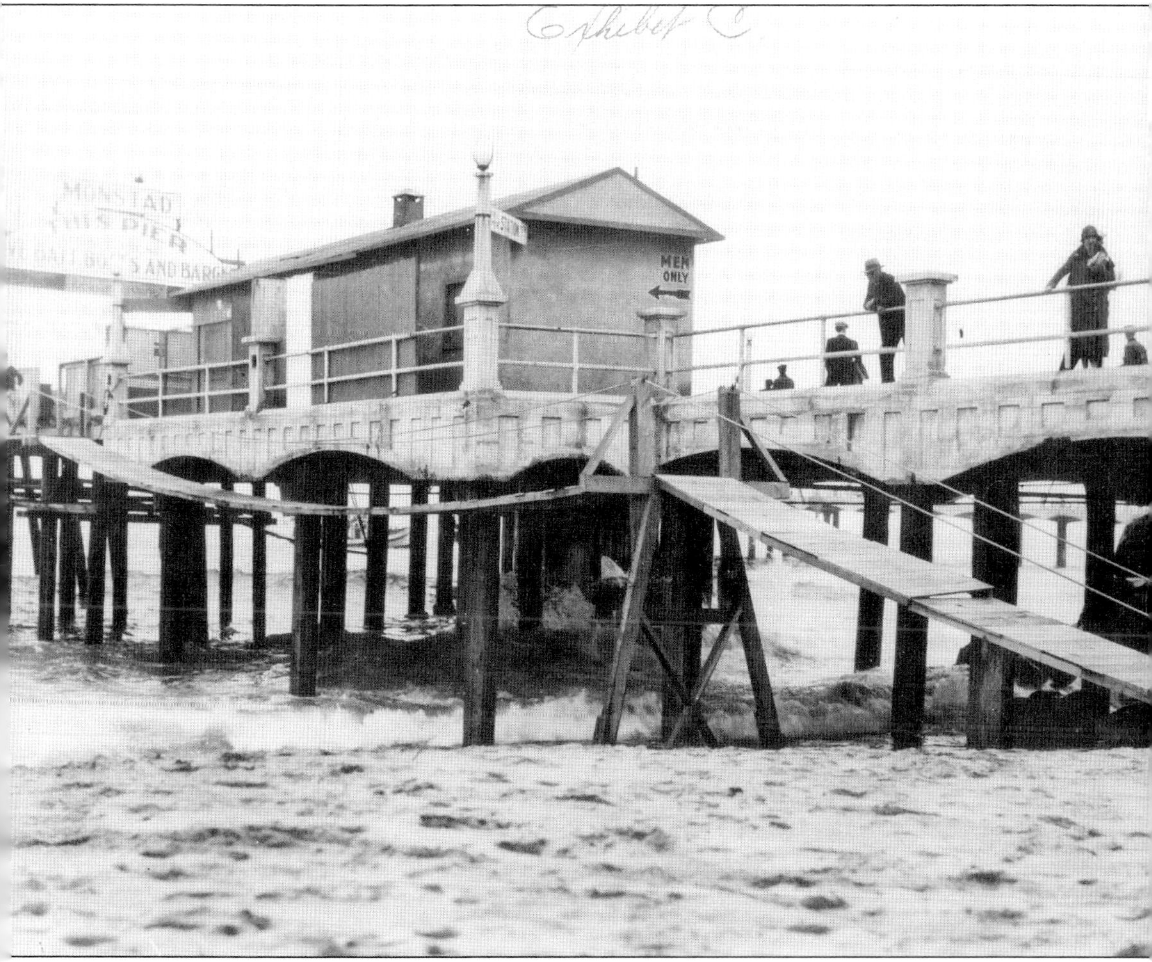

This is another view of the Monstad Pier taken around 1930. Today, it is still used for fishing, but no barges or boats load fishermen at this end of the pier. There are also several fresh fish vendors and picnic benches.

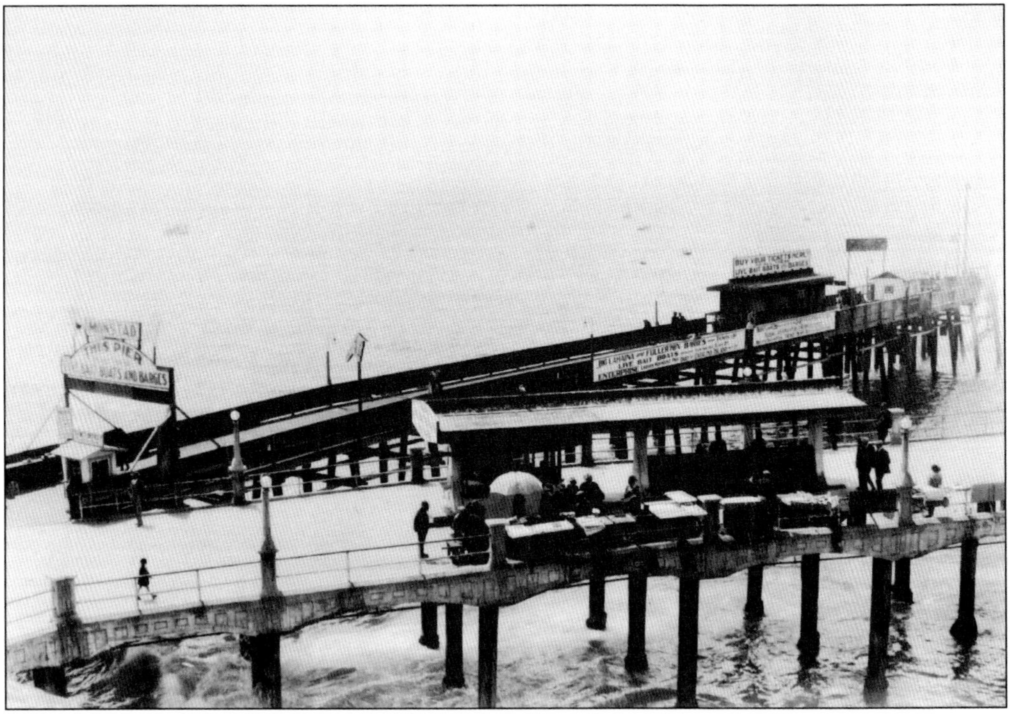

Monstad Pier sat adjacent to Horseshoe Pier. (Courtesy of the Redondo Beach Historical Society.)

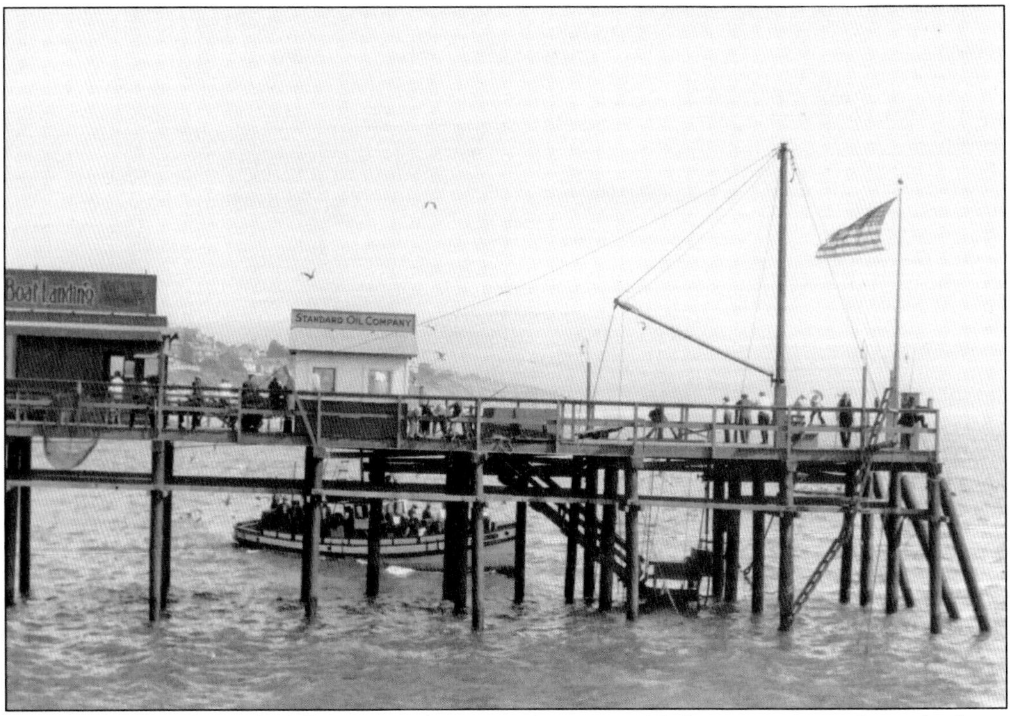

Pictured here at the end of the Monstad Pier is a boat most likely carrying people to the large fishing barges in the ocean. The Monstad Pier was called Fisherman's Wharf beginning in the 1940s or 1950s.

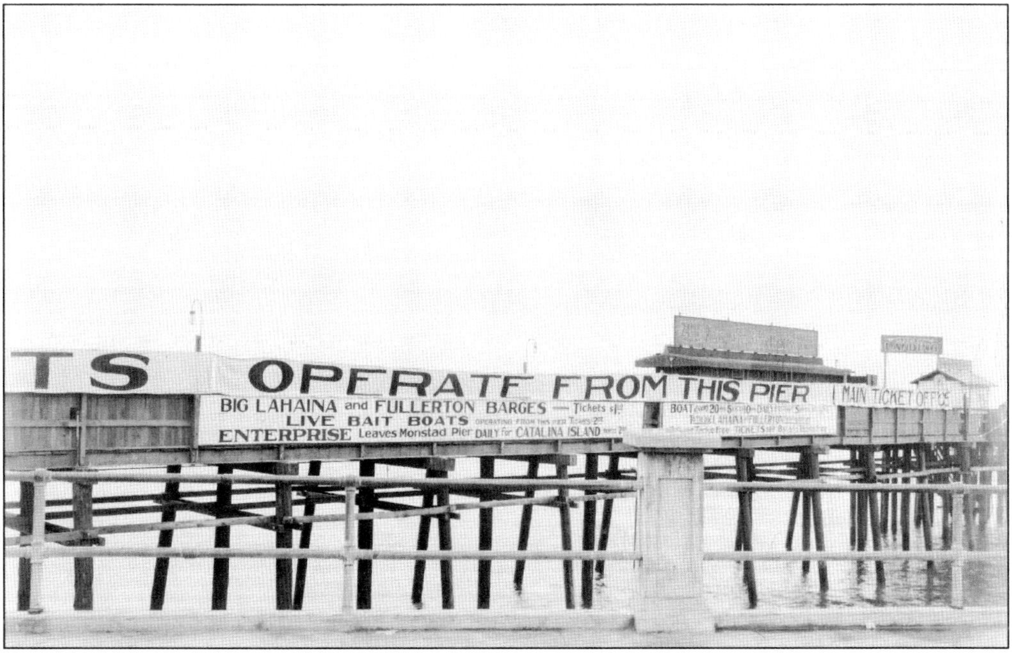

The Monstad Pier was used as a landing for fishermen who wanted to fish on large fish barges in the ocean. Boats left the pier every 20 minutes from 5 a.m. to 10 p.m. on a daily basis. A boat called the *Enterprise* took visitors to Catalina Island for $2. In the late 1930s, the pier and its operator, Capt. Webb Monstad, were under scrutiny by law officials because of an infamous illegal gambling ship, the SS *Rex*, that would anchor in the bay just far enough out to escape the grasp of law enforcement. Because the Monstad Pier was used as a landing, police were aware that boats could take people to the *Rex* illegally.

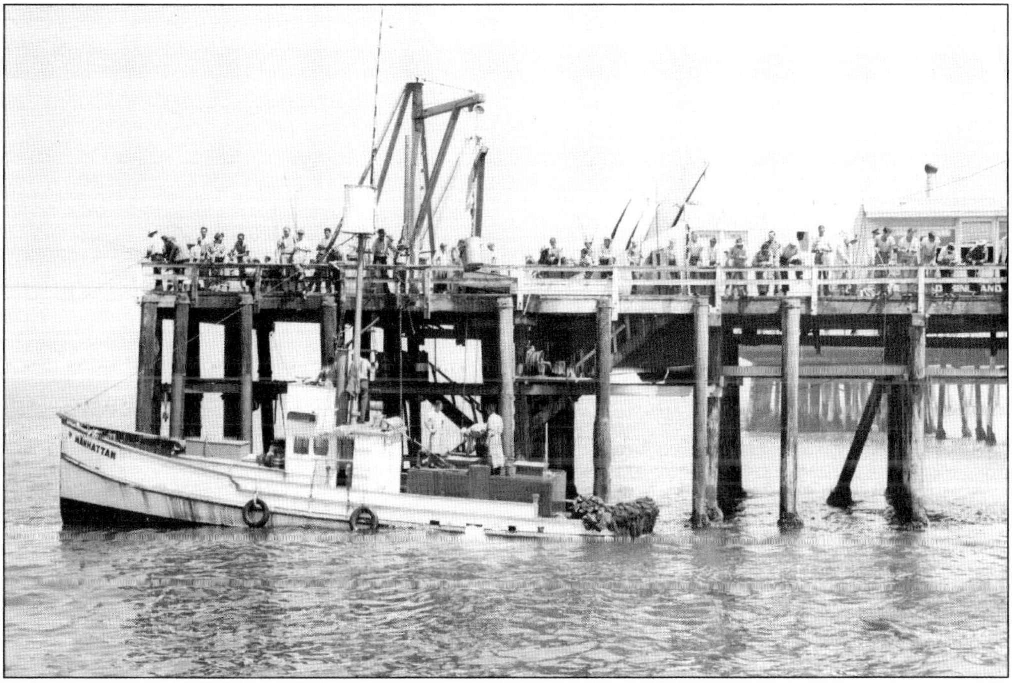

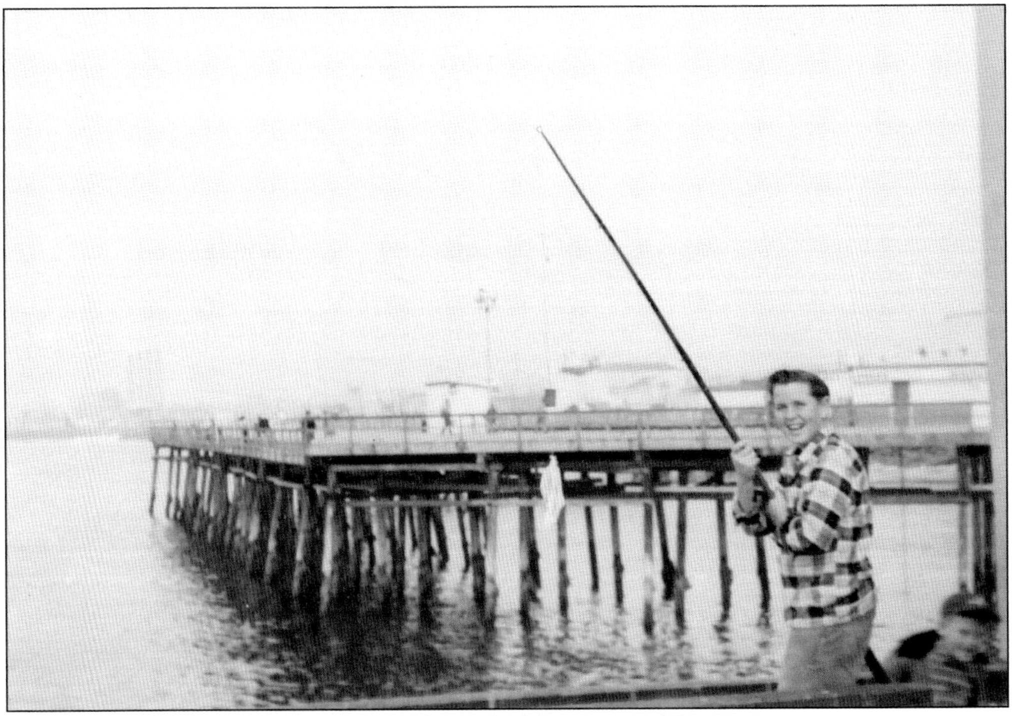

A young man in the 1950s fishes off of the end of the Monstad Pier.

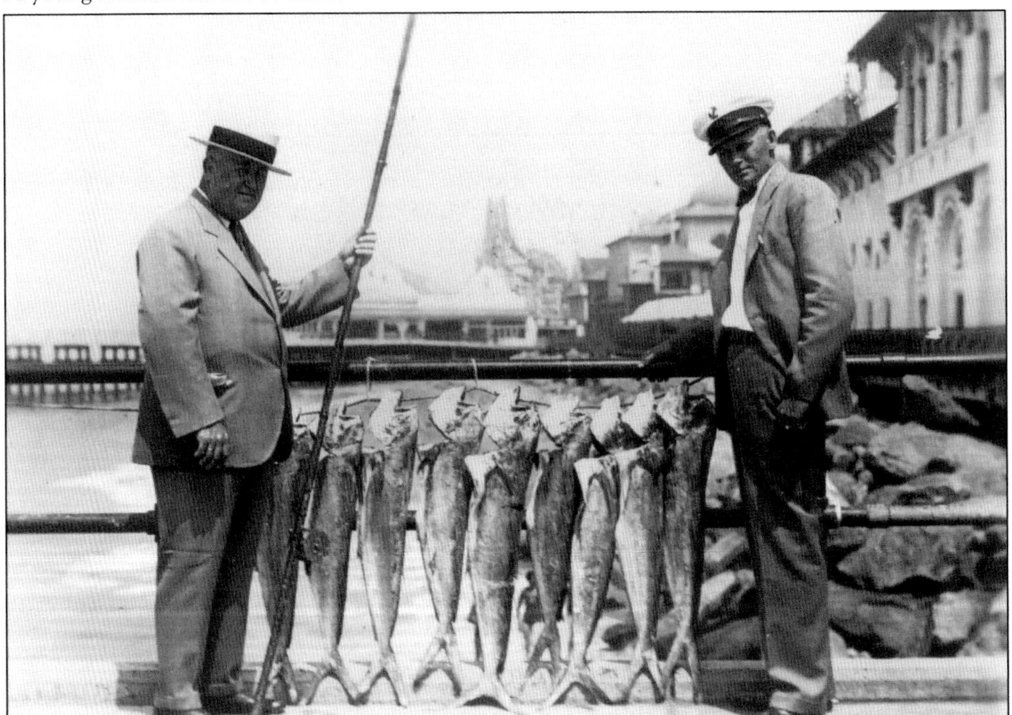

Marshal Albert C. Sittel of the Redondo Beach Police Department poses with Capt. Webb Monstad in 1930.

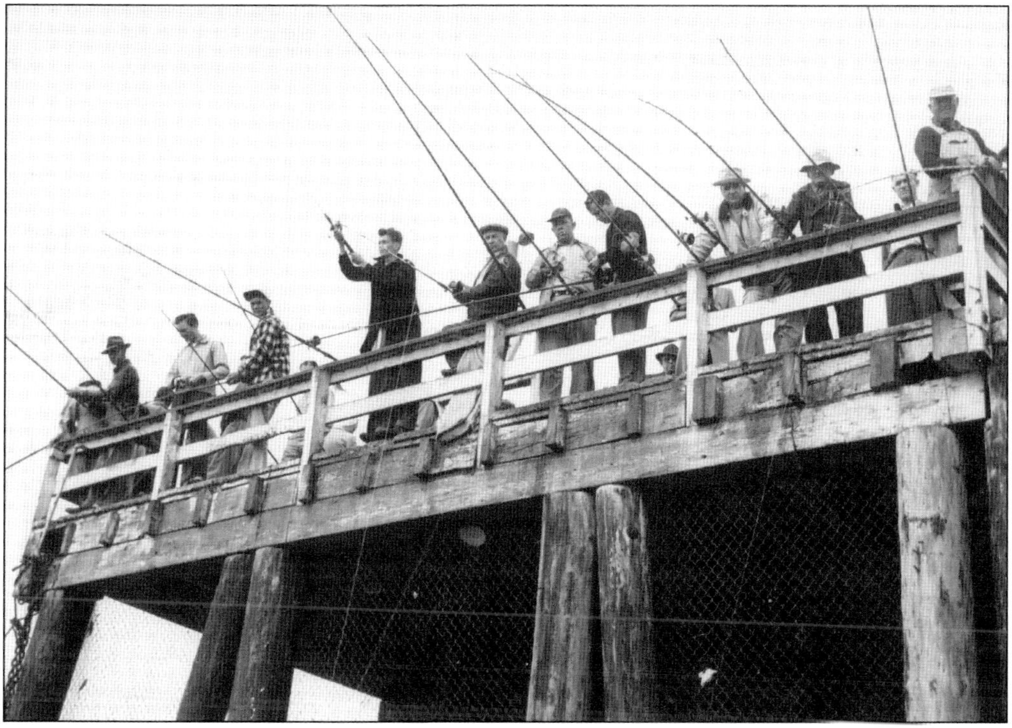

The Great Depression was difficult in Redondo as it was everywhere else. Not many people visited El Paseo or the piers to spend money. By the late 1930s, gambling circuits, including illegal ones, had set up along El Paseo. However, people were still interested in fishing.

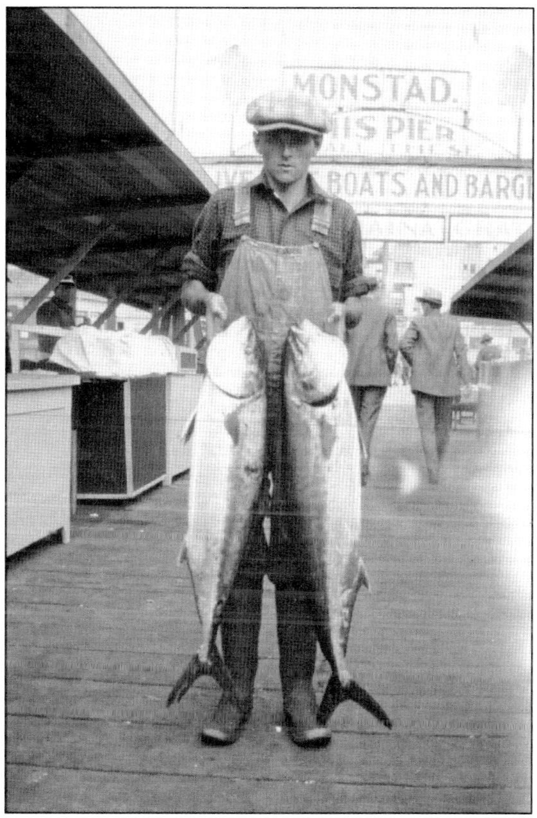

In this photograph from the 1930s, a young fisherman poses with two fresh yellowtail.

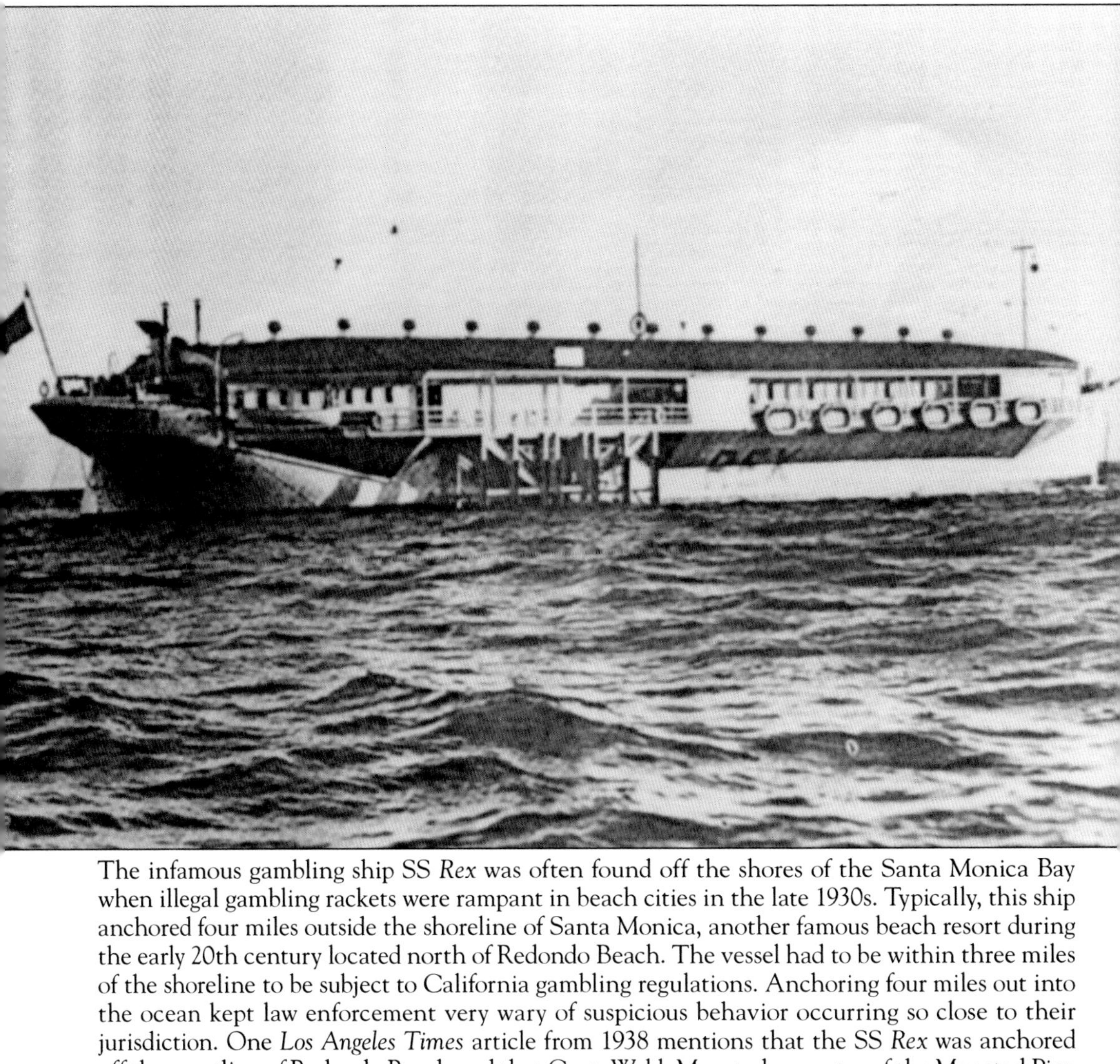

The infamous gambling ship SS *Rex* was often found off the shores of the Santa Monica Bay when illegal gambling rackets were rampant in beach cities in the late 1930s. Typically, this ship anchored four miles outside the shoreline of Santa Monica, another famous beach resort during the early 20th century located north of Redondo Beach. The vessel had to be within three miles of the shoreline to be subject to California gambling regulations. Anchoring four miles out into the ocean kept law enforcement very wary of suspicious behavior occurring so close to their jurisdiction. One *Los Angeles Times* article from 1938 mentions that the SS *Rex* was anchored off the coastline of Redondo Beach and that Capt. Webb Monstad, operator of the Monstad Pier, "declined to admit a connection" between his unusual appearance around the pier at the time of the equally unusual anchoring location of the *Rex*.

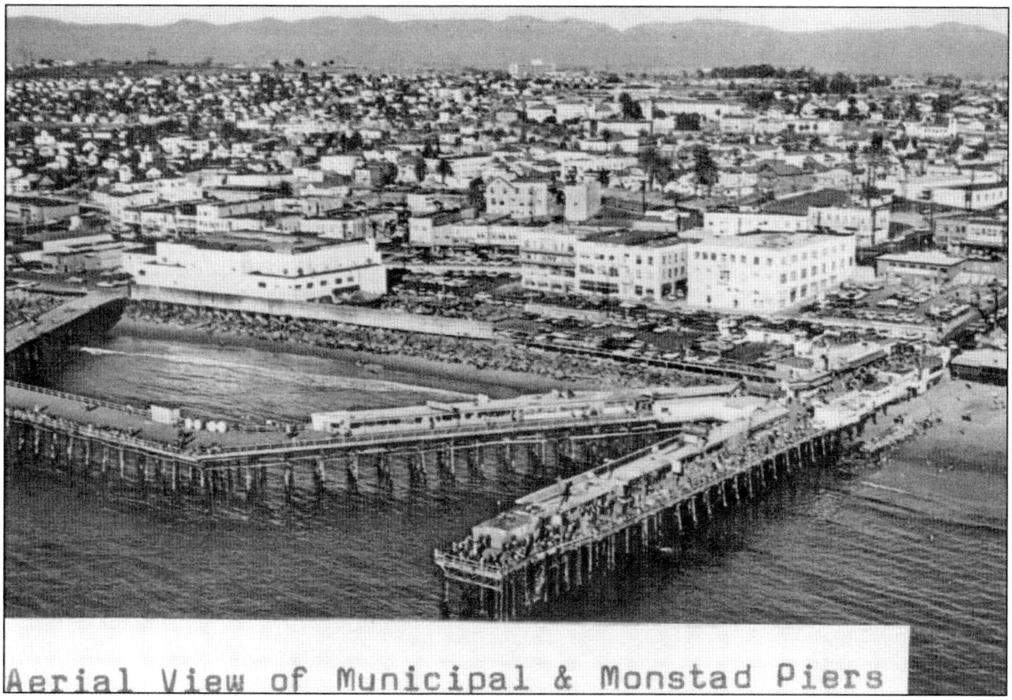

Although people were still coming to the pleasure pier, the area was in slow decline. The Pacific Electric Red Cars line was slowly pulling out of Redondo Beach. In 1922, the Pacific Electric sold to the city the beachfront north of the pier. In 1930, it sold the beachfront southward from the Bath House. Sometime between 1937 and 1941, the beachfront at the pier was sold to two separate syndicates. Above is an aerial shot of the piers after 1941. The Plunge was demolished in 1941 because of a decline in patronage and a rise in operating costs. Below is a postcard from around the same time.

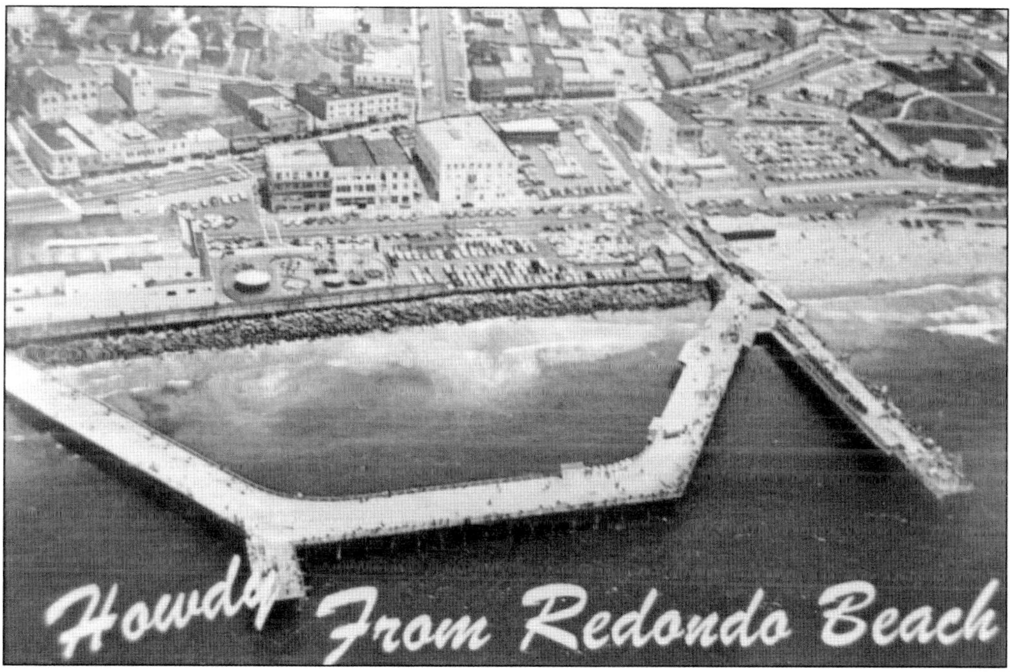

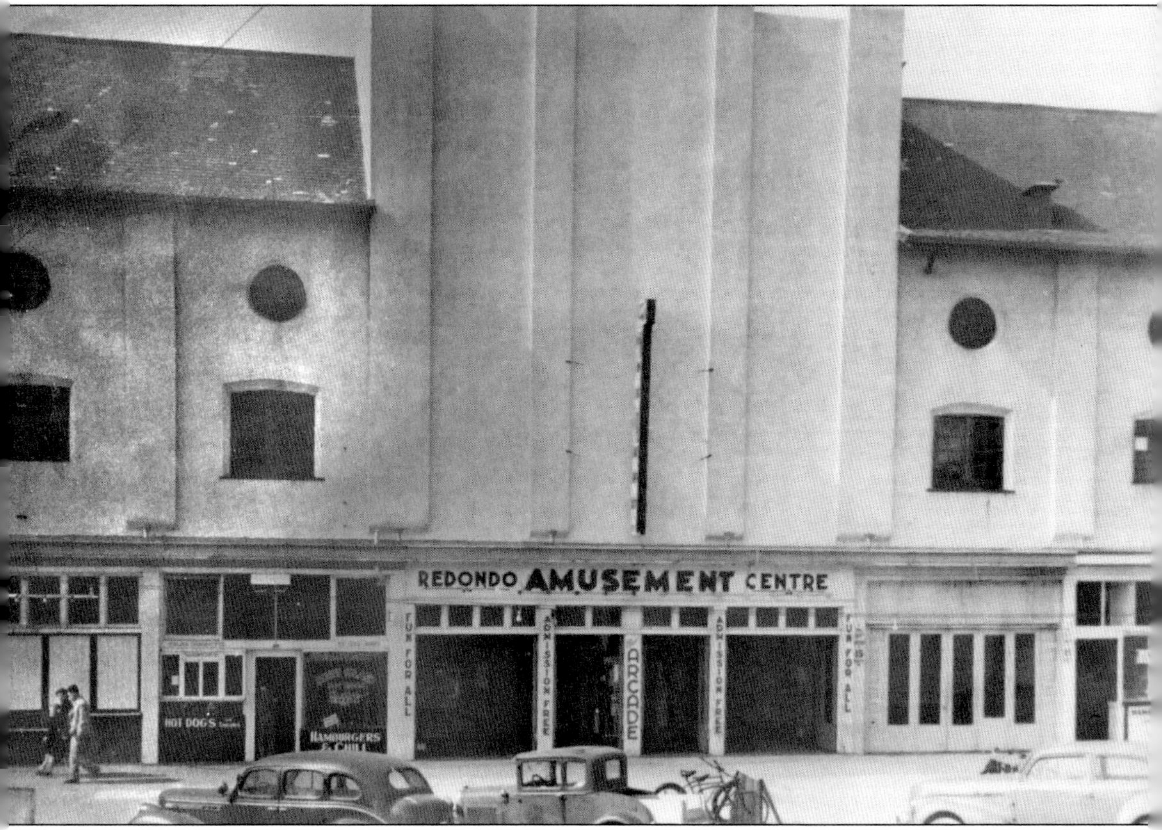

After the Plunge was demolished, the vacant lot where the building once stood was redeveloped into a parking lot. The Pavilion was given an updated facade that was stylistically similar to late Art Deco architecture. This is a photograph of the entrance to the Redondo Amusement Centre at the Pavilion.

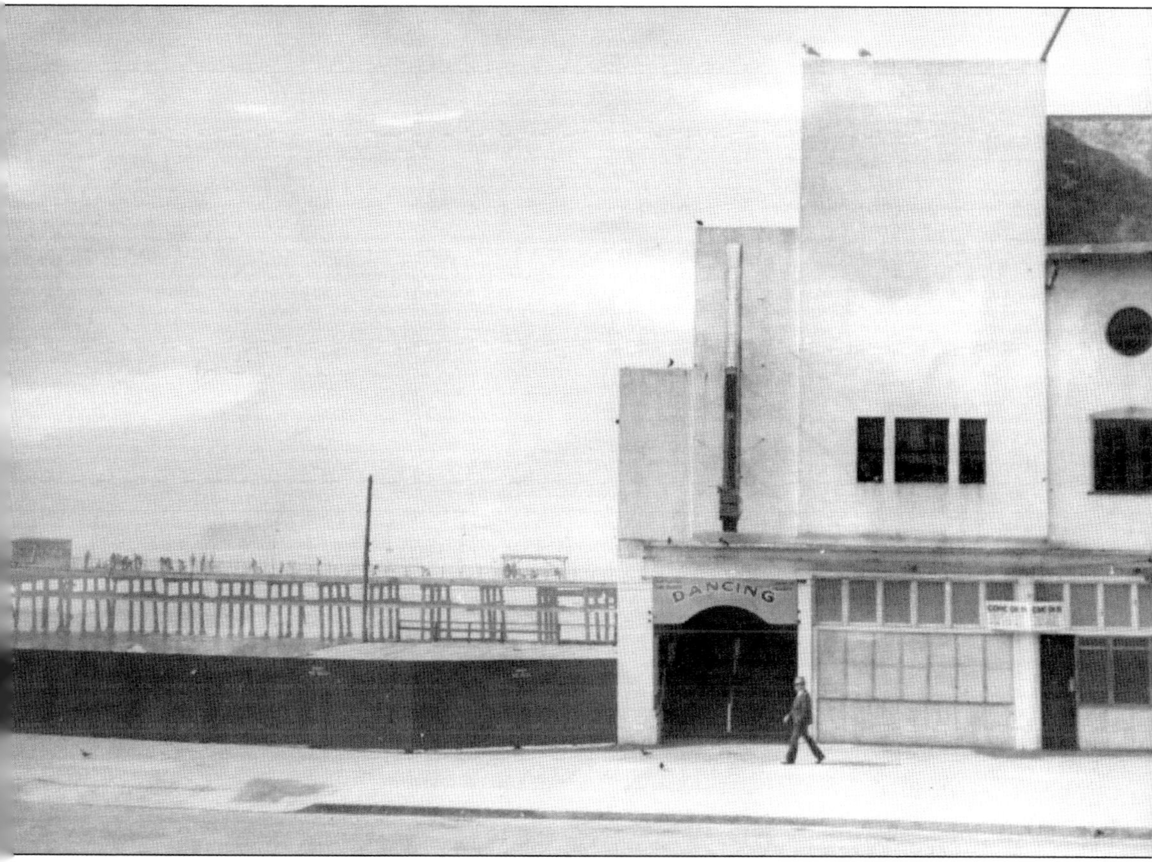

In the image above, a man walks past the stairs leading to a dancing facility. After the height of the pleasure pier era, dancing was still an option at the pier.

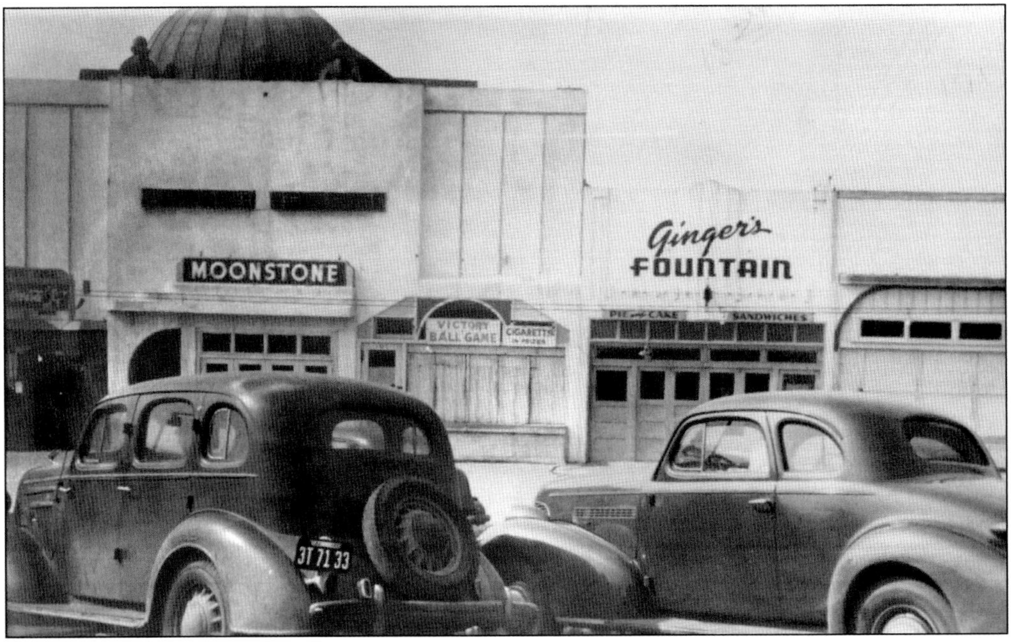

Along El Paseo, buildings received the same architectural treatment as the Pavilion in the 1940s. Pictured here are businesses along El Paseo.

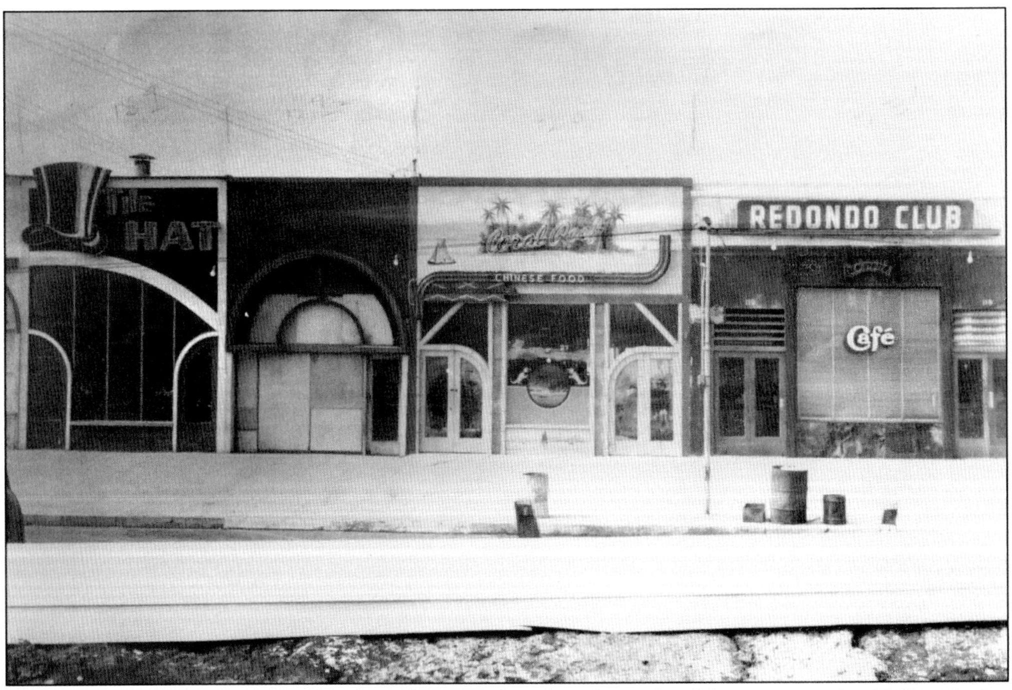

Above is a strip of storefronts that have been updated to the style of the times in the 1940s. To the right of The Hat and Coral Reef Chinese Food is the Redondo Club, which was a bar and café.

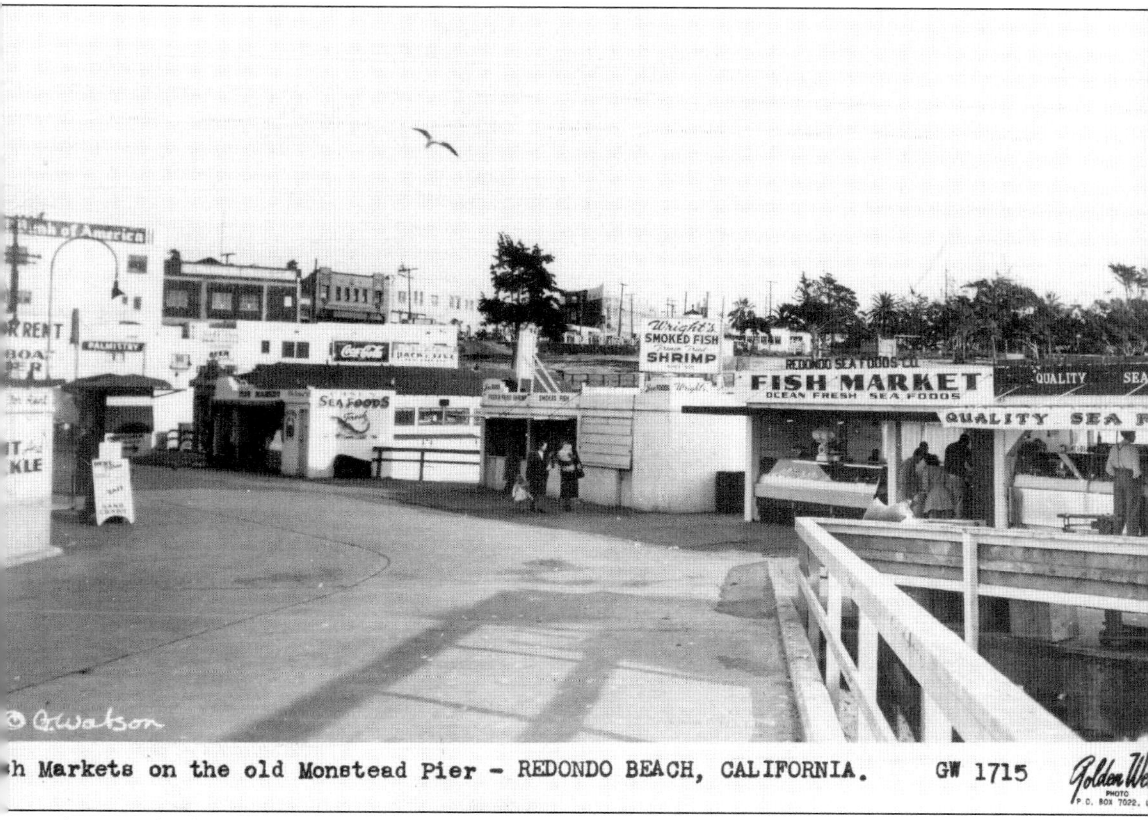

Photographed above is the Monstad Pier in the 1950s. By this time several more fish markets and seafood kiosks had opened up on the pier.

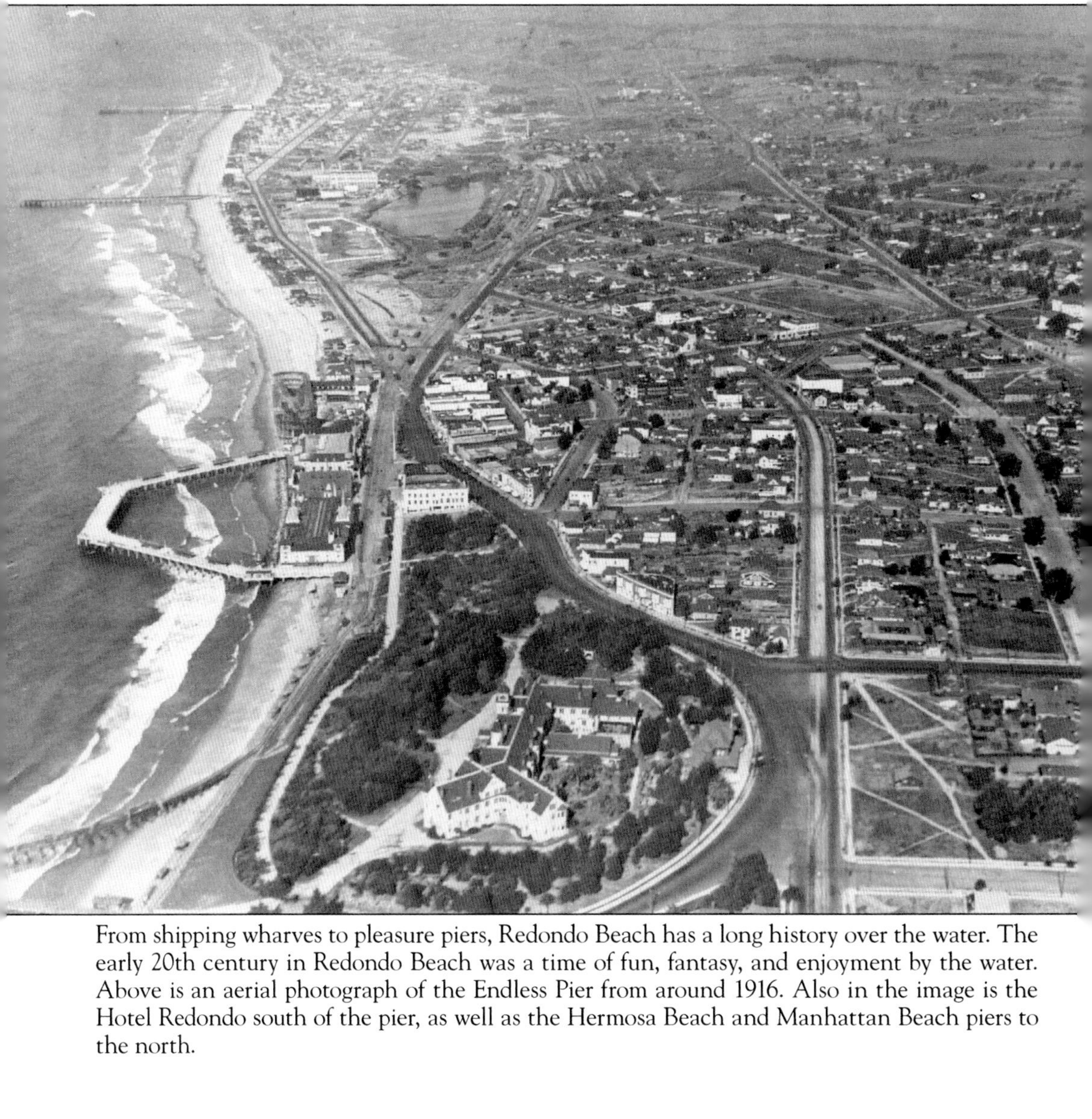

From shipping wharves to pleasure piers, Redondo Beach has a long history over the water. The early 20th century in Redondo Beach was a time of fun, fantasy, and enjoyment by the water. Above is an aerial photograph of the Endless Pier from around 1916. Also in the image is the Hotel Redondo south of the pier, as well as the Hermosa Beach and Manhattan Beach piers to the north.

Three

THE ARCHITECTURE OF FUN AND FANTASY

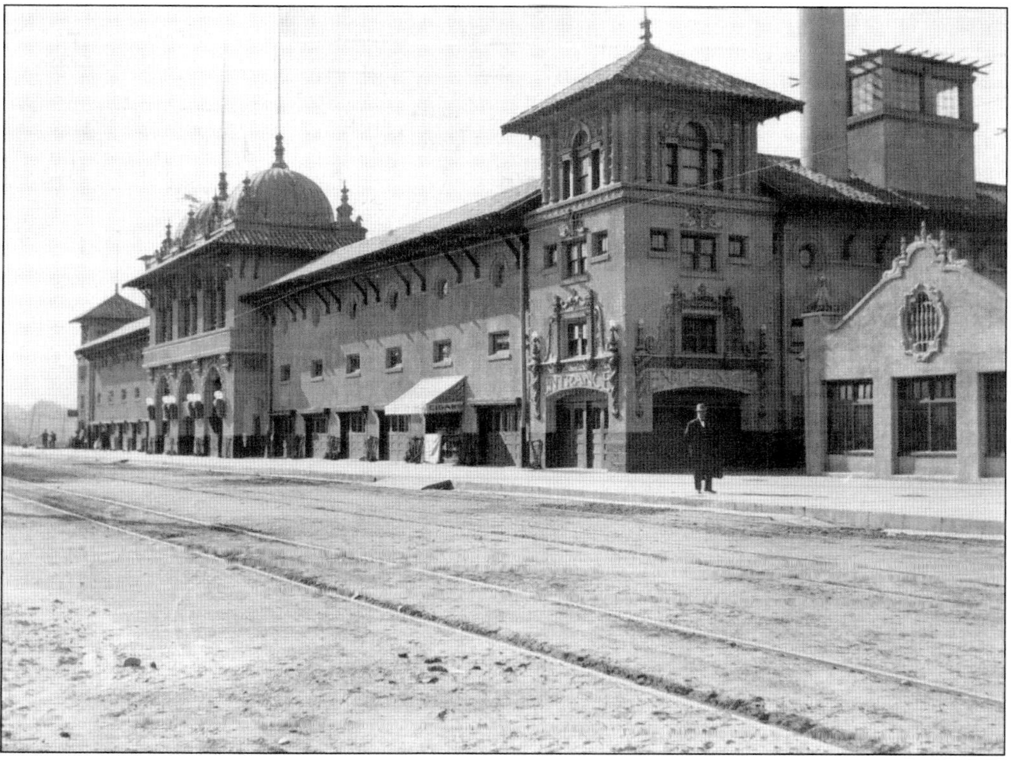

The shoreline of Redondo Beach in the early 20th century reflected that of a fantasyland on the water. Stylistically, pier buildings flaunted whimsical, pleasurable, and engaging architectural design. Flags, towers, and turrets adorned many of them. The idea of a pleasure pier was reflected in the design of these piers.

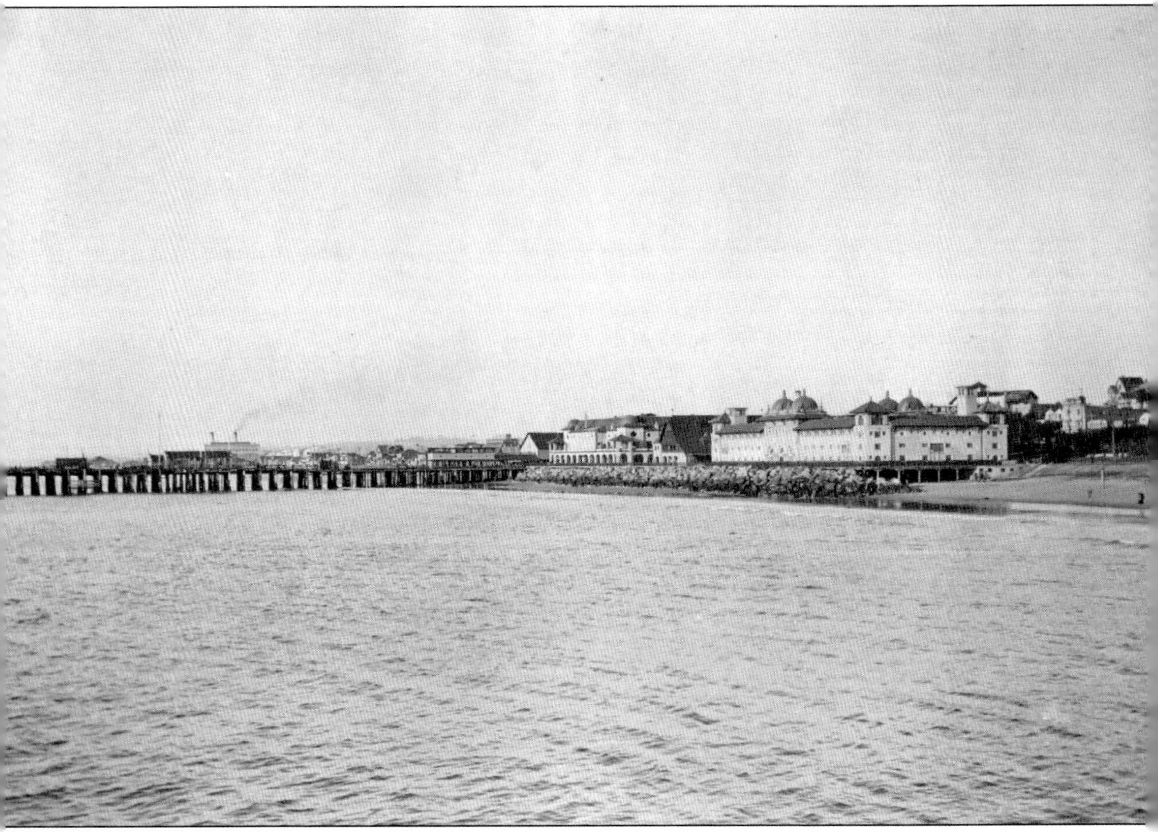

Several types of buildings were found on or near the pleasure pier. Hotels, bathhouses, and pavilions were the notable standard structures associated with pleasure piers and the main buildings of beach resort towns.

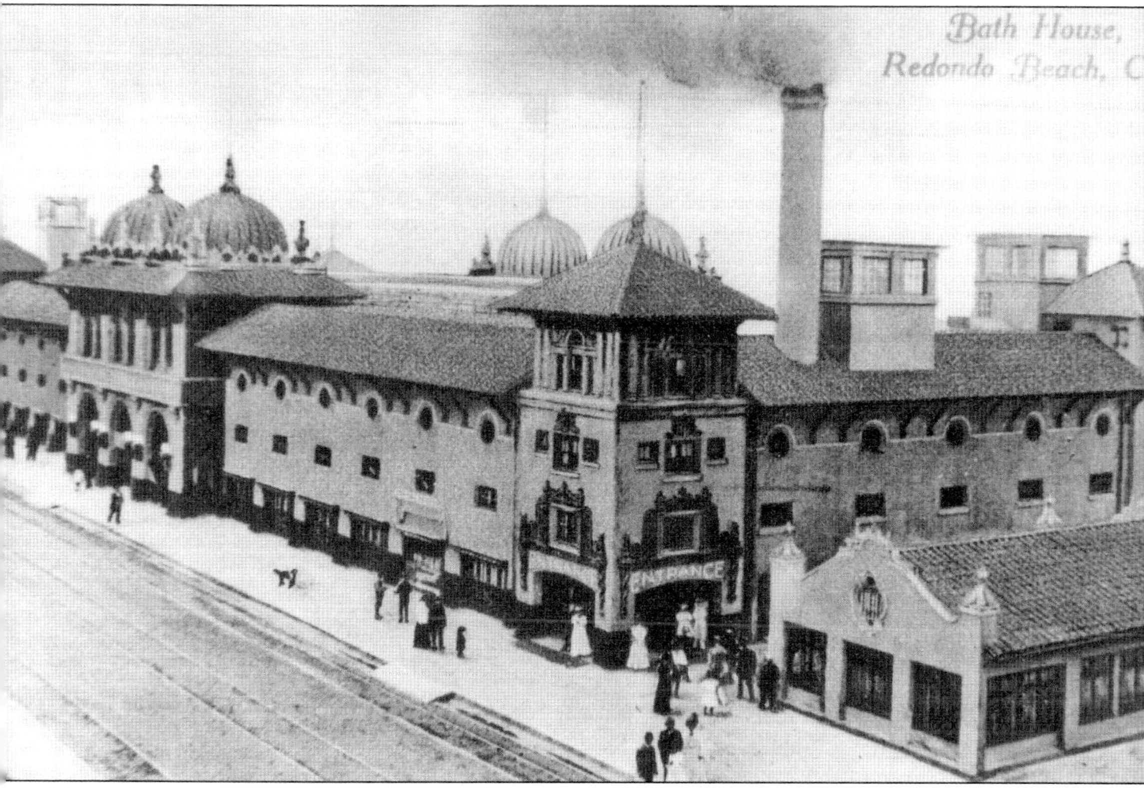

The architecture of these buildings evolved as the attitude of beachgoers changed from the modest Victorian era and mind set of the late 19th century to more exotic styles that conveyed a message less elite and more pleasurable.

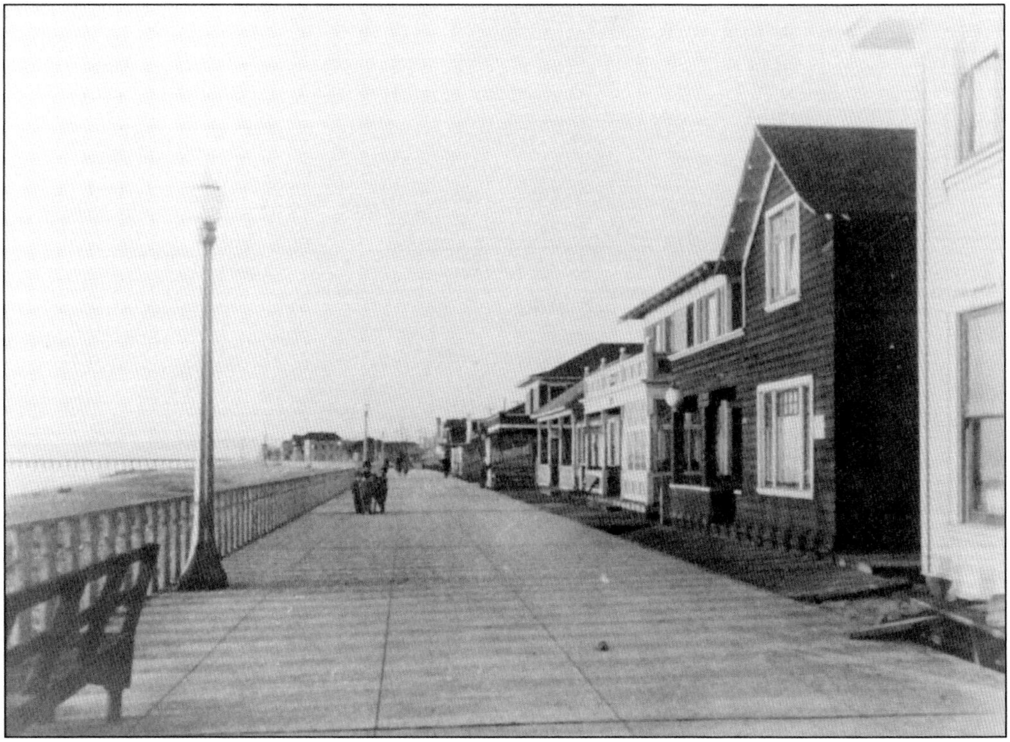

During the late 19th century, Redondo's architecture was not as grandiose or ostentatious. It represented a time in pleasure pier architecture when modesty was of the utmost virtue. The upper echelons of society resorted at the beach and attempted to provide education to the masses through Chautauqua assemblies. The architecture reflected this elite attitude toward how to vacation at the beach. Victorian and Neoclassical architectural elements were used in upper-class residential houses. The manner of architecture indicated who dictated the beach atmosphere in a given area.

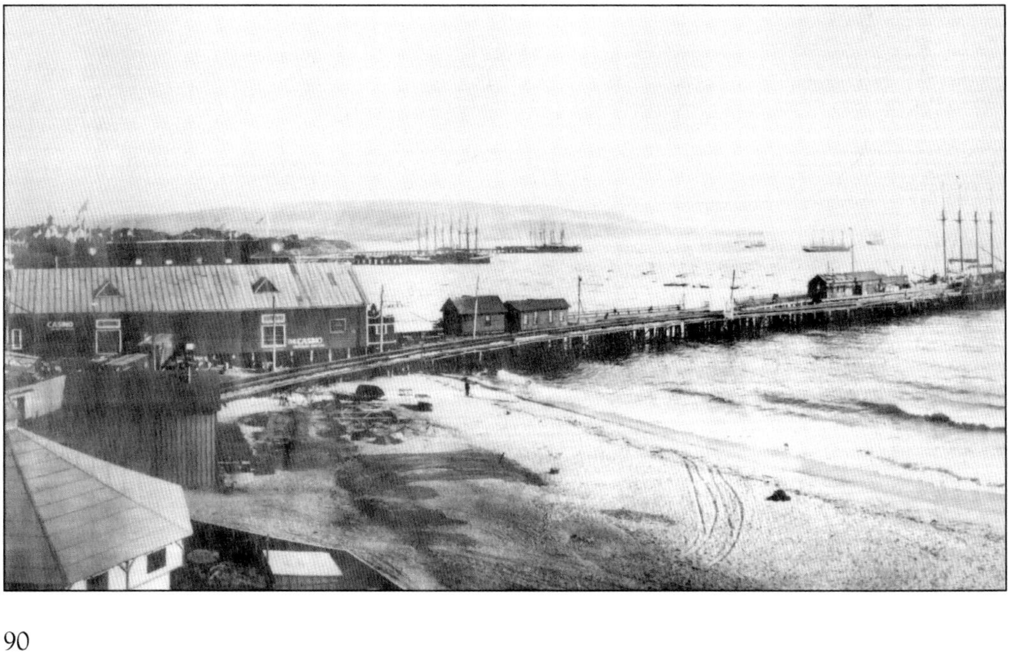

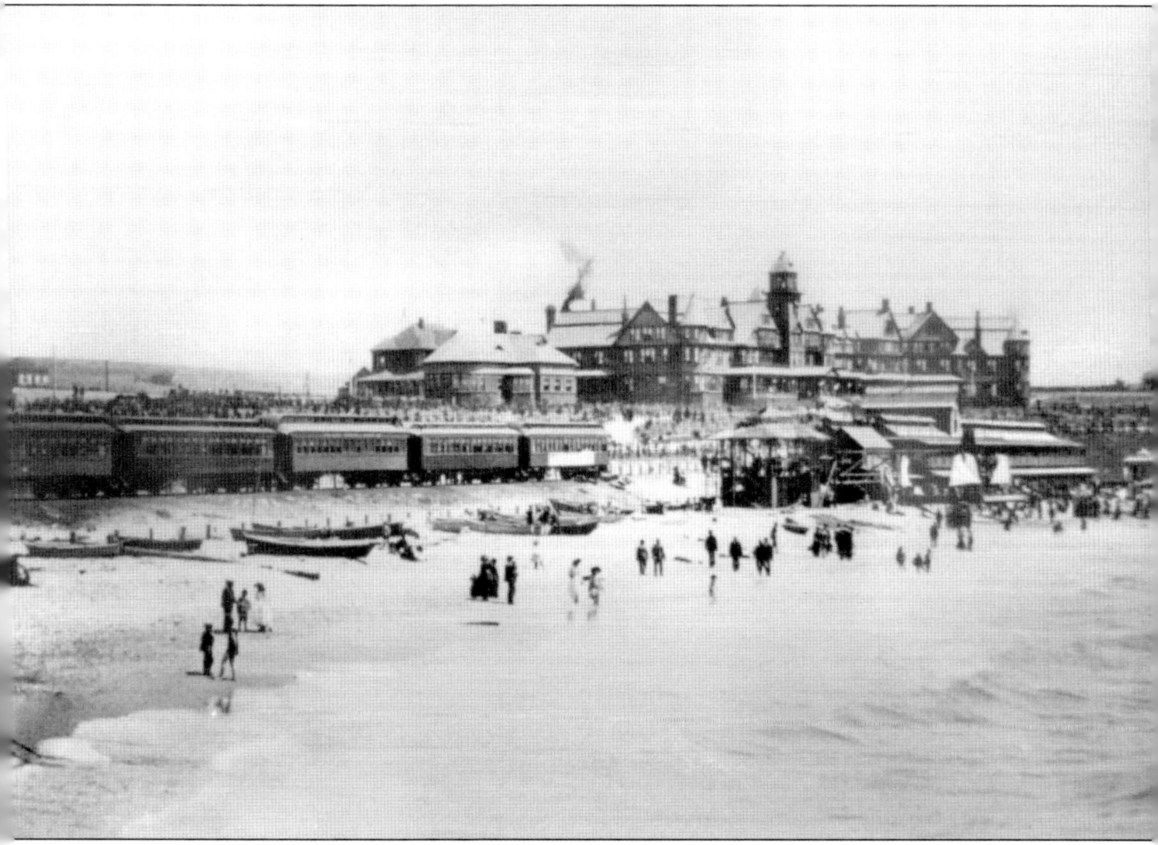

Modesty was still a virtue, even at the beach. Bathhouses had separate baths for men and women. These values reflected the Victorian mind set of the early 1900s. As such, the early pleasure piers were still places of learning and natural enchantment.

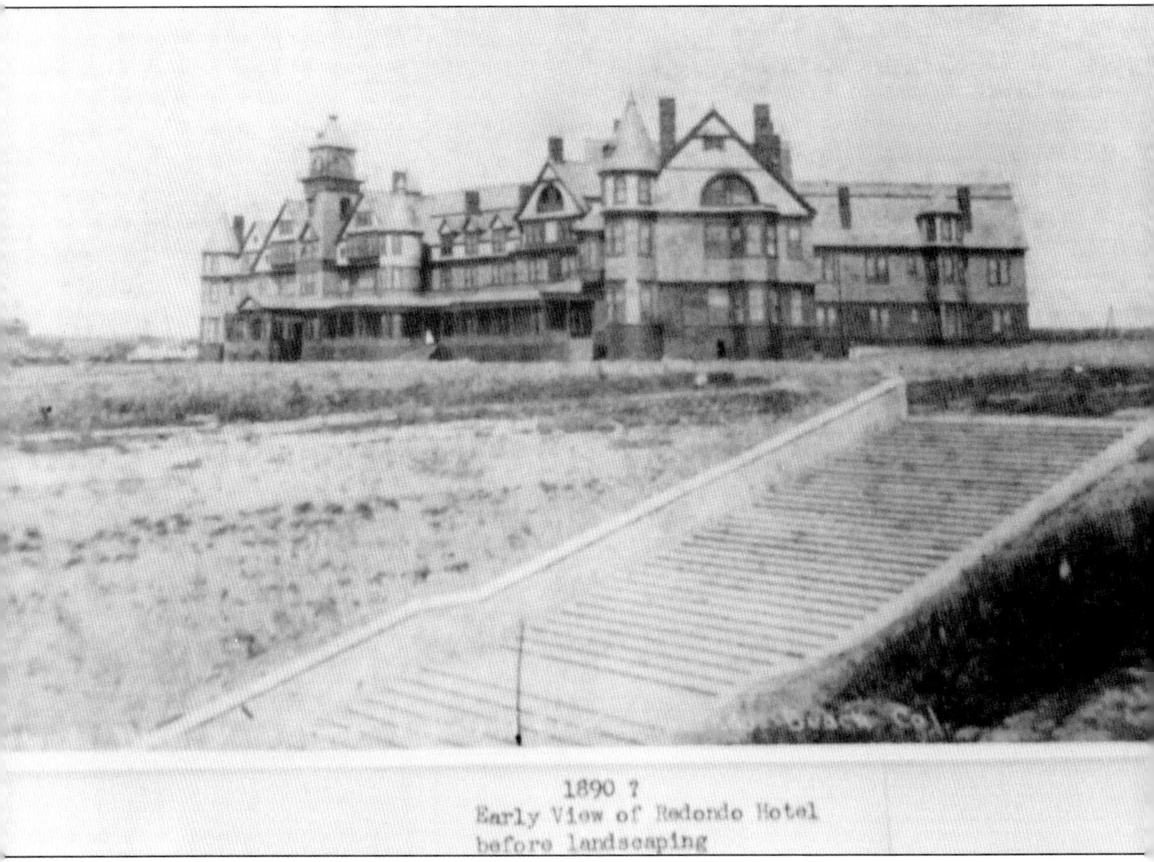

An example of early Victorian architecture is the Hotel Redondo. Constructed in 1890, it stood as a beacon of elegance, beauty, and class. It was not cheap to sleep there. Eventually, the hotel fell out of style, and it was demolished in 1925. Its wood was salvaged for other projects. Part of the staircase in this picture remains today.

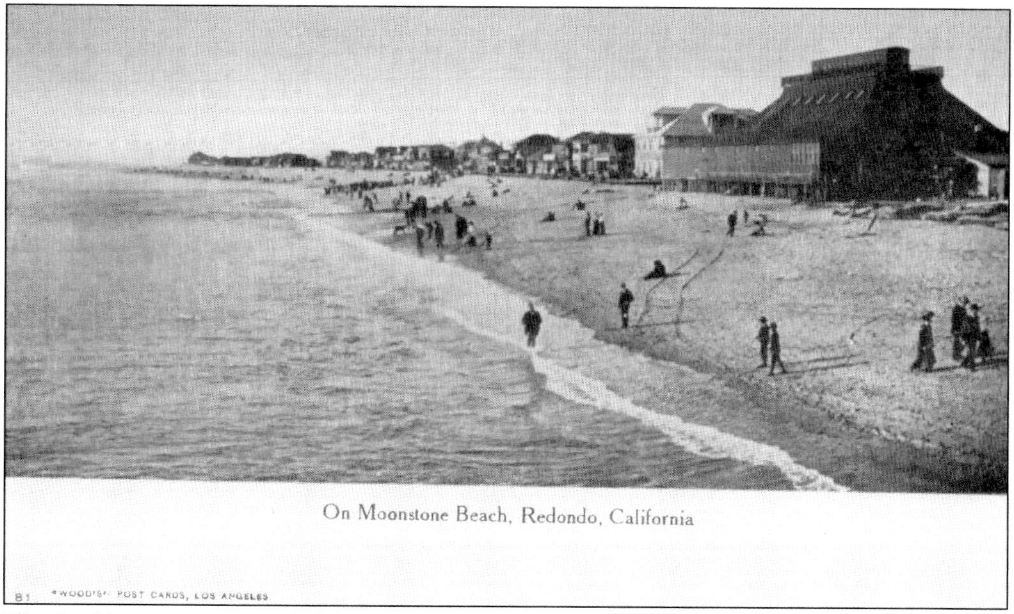

The buildings constructed during the late 19th century were typically modest in size and ornamentation. They were usually wood-frame buildings with clapboard siding. Here is a picture from around 1900 of the buildings at Moonstone Beach just north of where the pleasure piers were eventually constructed years later. (Courtesy of Donald Krintz.)

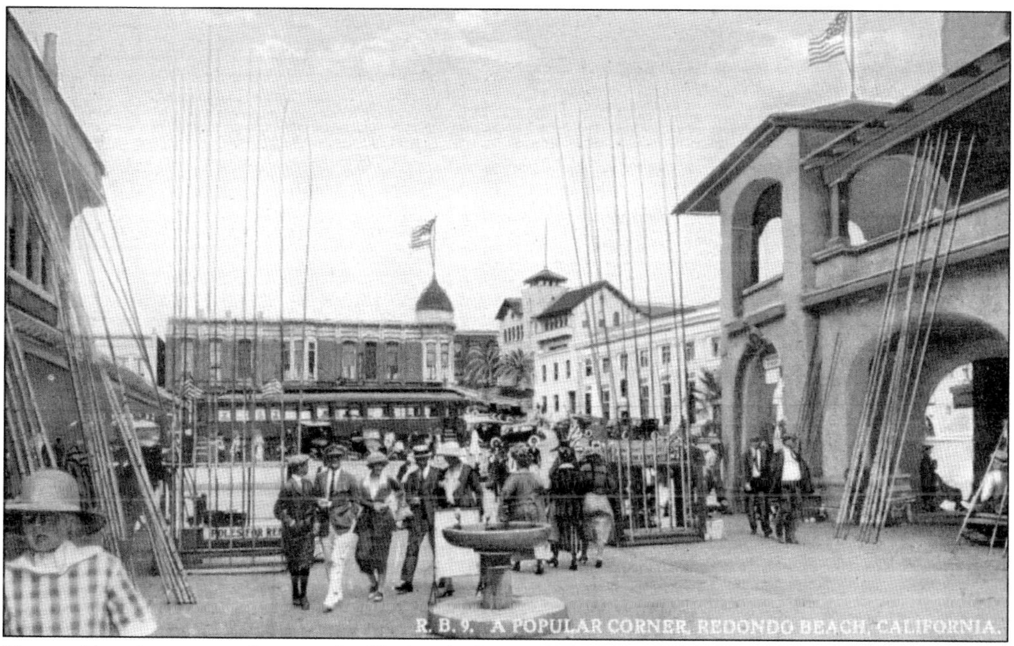

Henry Huntington took over the Redondo Improvement Company in 1905. It was under his direction that larger, lavish, more-permanent buildings were constructed to bring in tourists, financiers, potential homeowners, and commercial investors. The architecture he introduced to Redondo's shoreline was much more robust and whimsical. It was influenced by foreign styles that made coming to the beach much like going to a fantasyland.

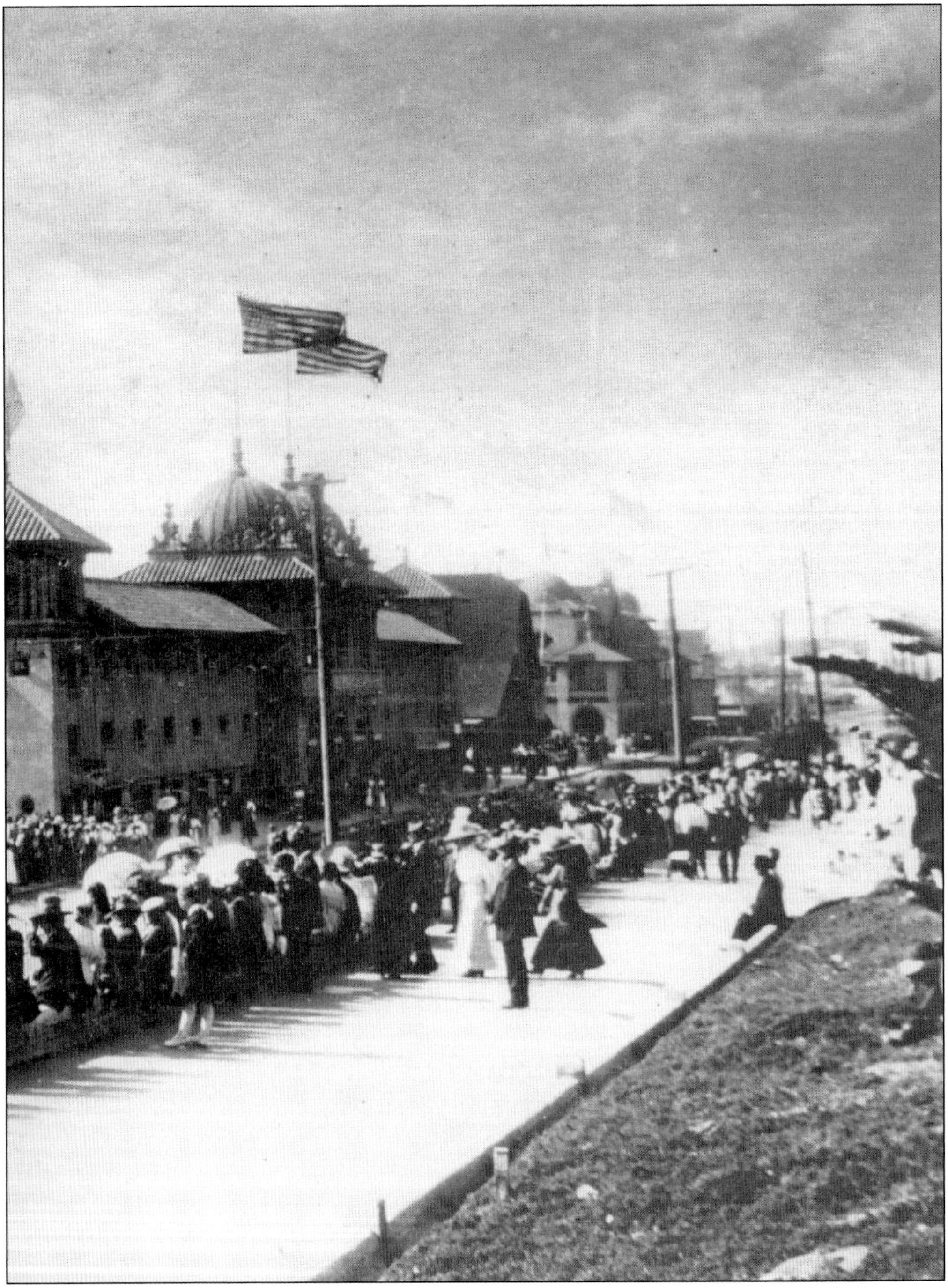

The American Exotic Revival style was influenced by several foreign styles, particularly Egyptian and Moorish architecture. This architecture was typically used in hotels, theaters, and garden pavilions—in short, places of pleasure. It is no surprise that Redondo Beach's pleasure piers embraced this architecture in the early 20th century for its seaside architecture. This photograph is of El Paseo some time around 1900. The Plunge, the casino, and the Pavilion can all be seen on the left.

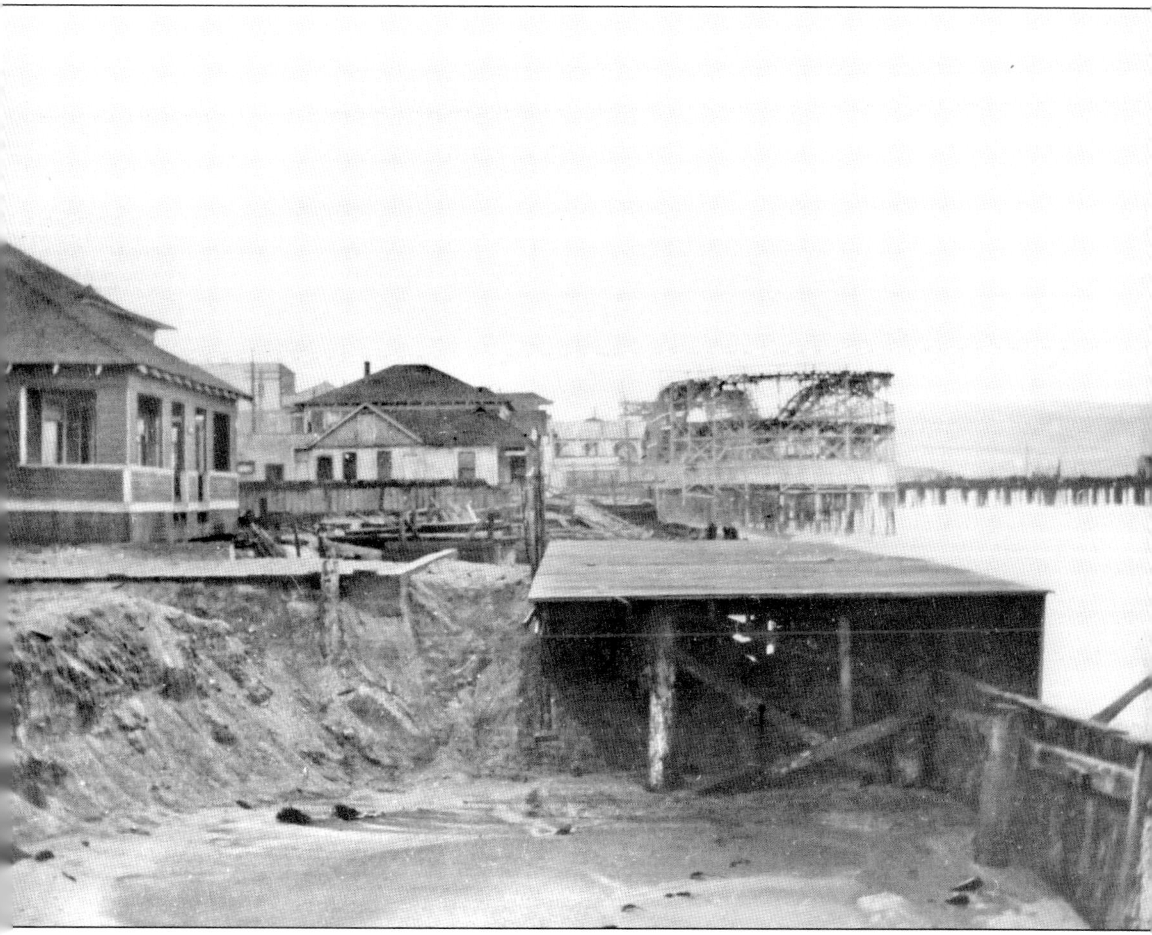

This photograph was taken after the storm of 1915. The damaged Lightning Racer roller coaster can be seen in the background.

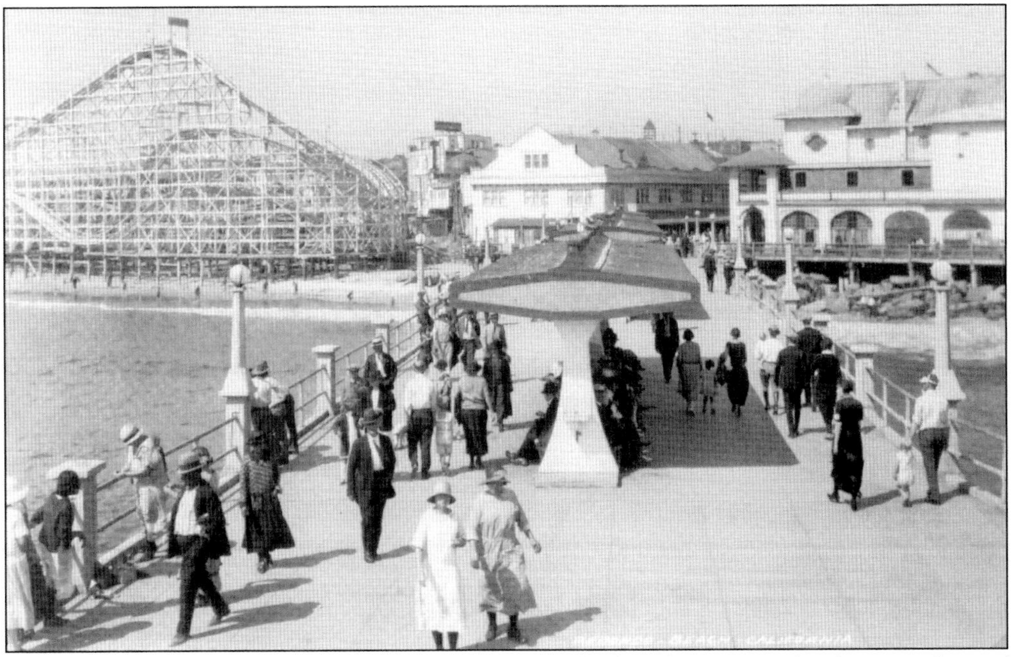

By the early 1900s, the mentality and ambiance of the beach resort evolved into a place of fantasy and fun. Catering more to the burgeoning middle class of America, the next generation of architecture expressed notions of exotic enchantment, international wonder, and overall, a philosophy of pure American fun. The pleasure piers had changed from places of the elite to amusement parks that appealed to the masses. Above is an image of the northern leg of the Endless Pier and Giant Dipper roller coaster about 1920. The first roller coaster was the Lightning Racer or Racing Coaster. It was destroyed in the storm of 1915 and was replaced with the Giant Dipper.

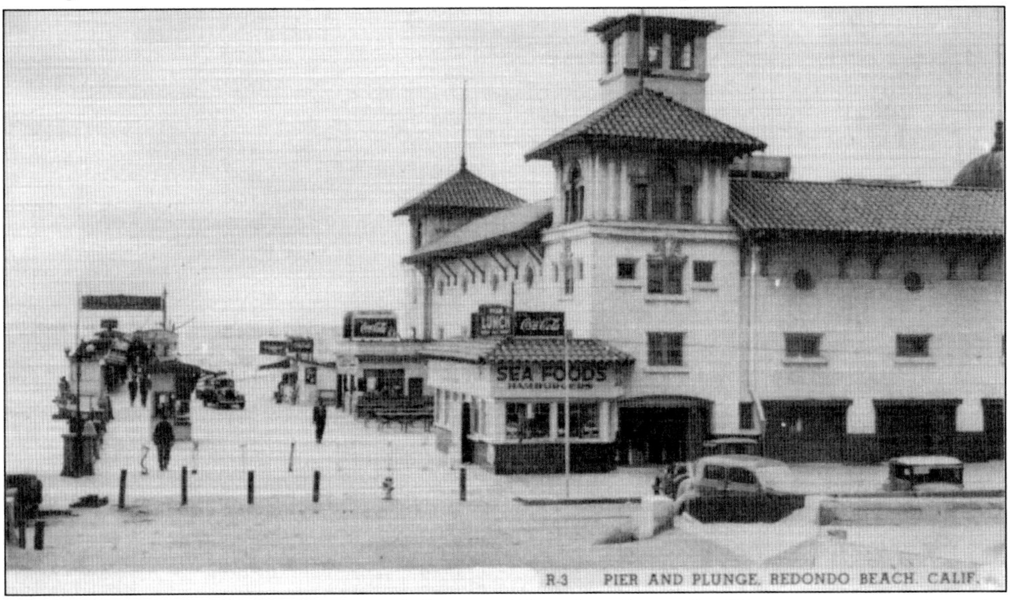

Above is a postcard of the Plunge and the pier. Note how large the Plunge is compared to the entertainment venues around it. These grand buildings were typically constructed at a palatial scale to cater to the fantasyland atmosphere that was created at the beach.

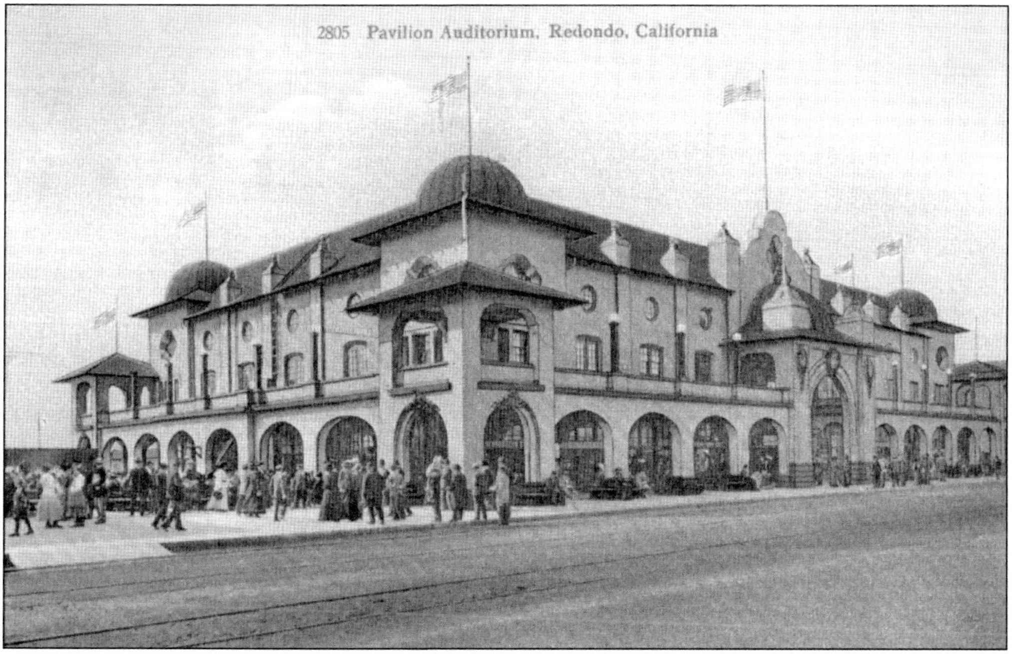

The first building that Henry Huntington constructed to attract tourists to Redondo was the Redondo Pavilion. Constructed in 1907, the main feature of this building was the Mandarin Ballroom, which seated up to 4,000 people. According to a *Los Angeles Times* article, the ballroom was on the second story and "featured a beechwood dance floor with space for 500 couples." On the first floor were shops, a theater, and a restaurant. The Pavilion was a 34,069-square-foot structure, built three stories high in the Mission Revival style. The wood-frame structure was clad in white stucco. The roof was covered in red clay tiles and was ornamented with finials.

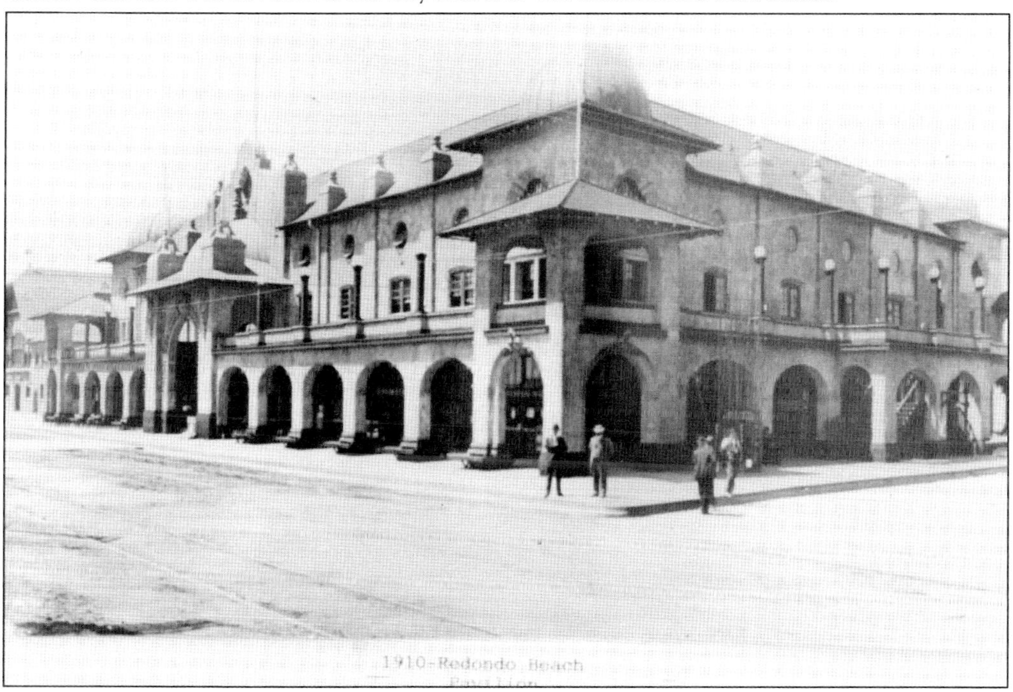

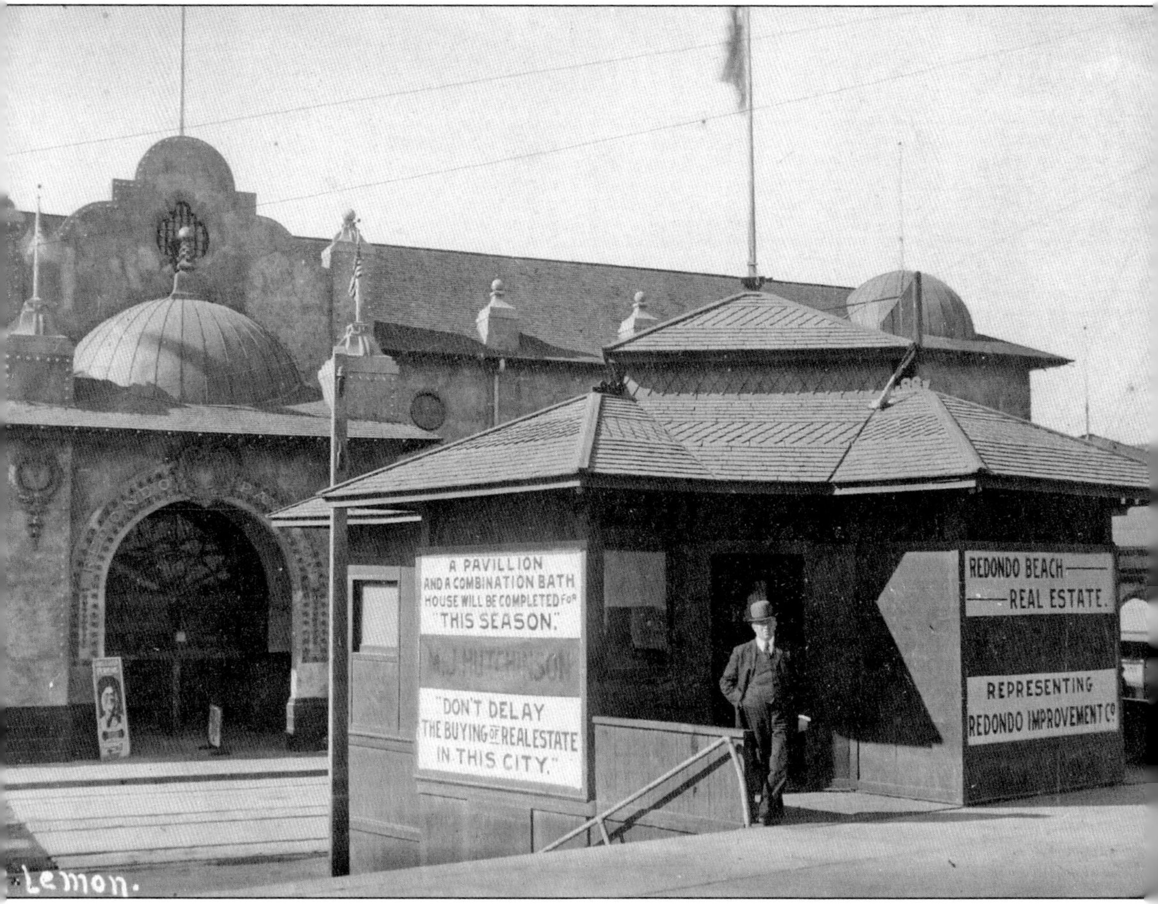

This is a picture of a small kiosk in front of the grand entrance to the Redondo Pavilion. The sign on the kiosk reads, "A Pavilion and a Combination Bath House Will Be Completed This Season. Don't Delay The Buying of Real Estate in This City." Another sign reads, "Representing the Redondo Improvement Co." Obviously, the Pavilion and the Bath House were constructed to entice potential homebuyers and commercial developers to Redondo Beach.

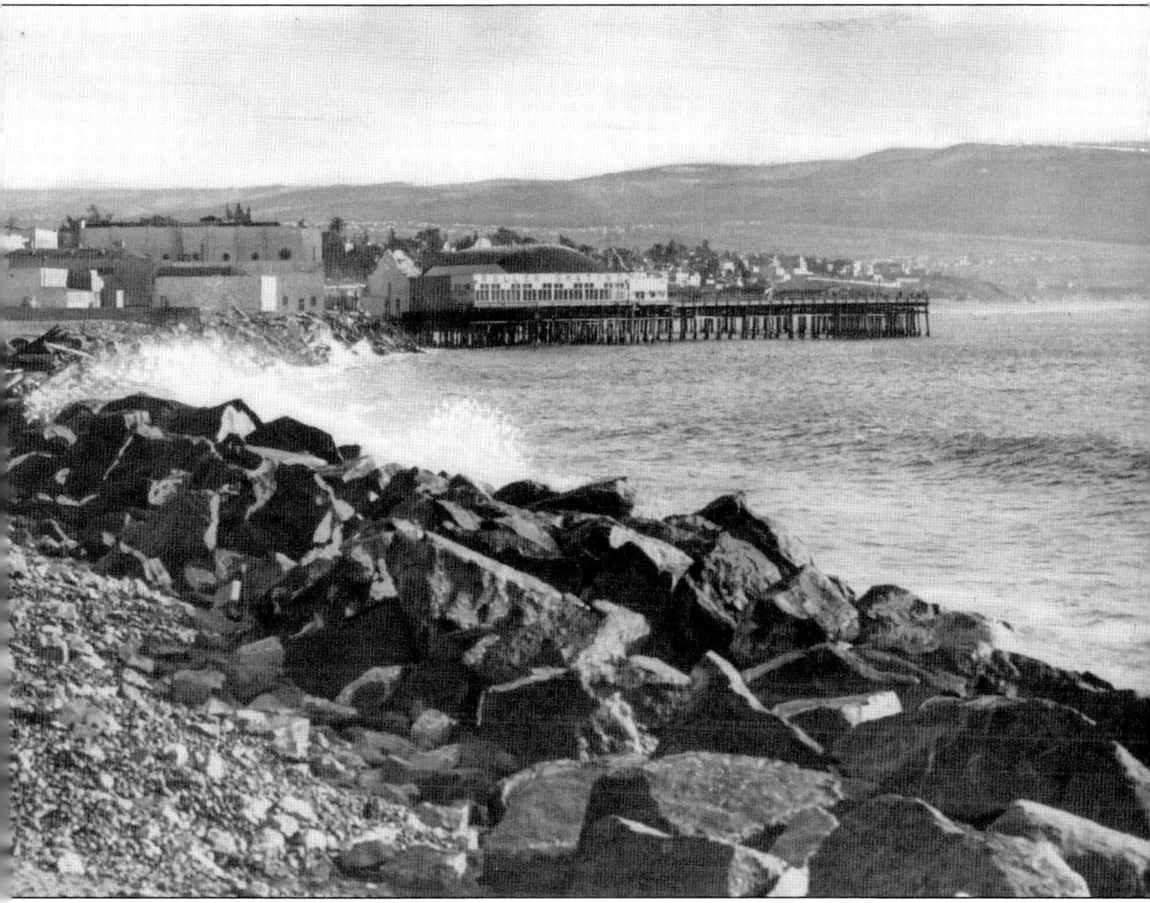

The pier and the Pavilion are seen in this 1940s photograph. The Plunge is gone, but there is a roller-skating rink. The Pavilion has been given a new facade that is more up to date with the Art Deco architectural style of the times.

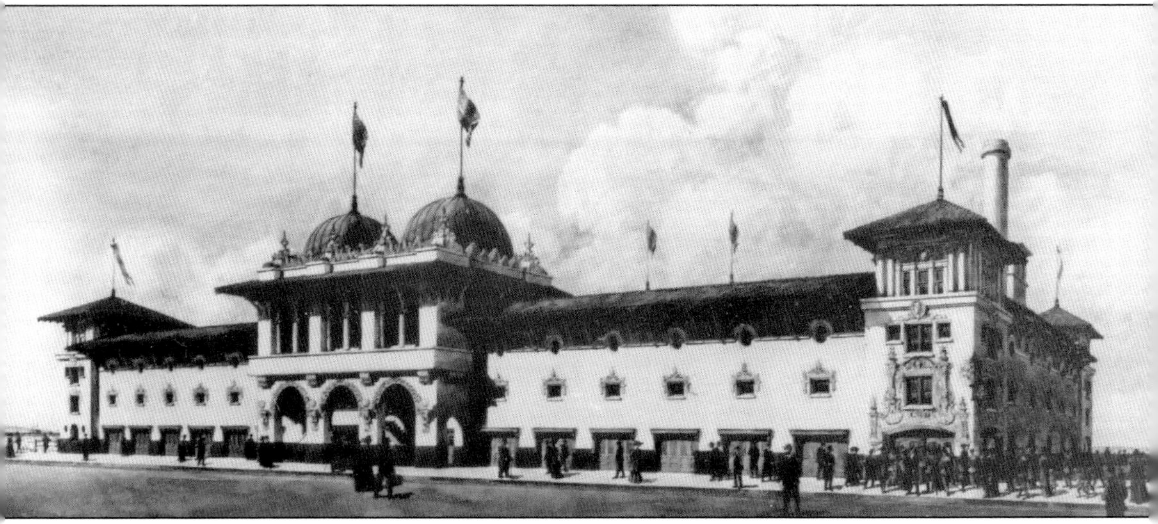

The Redondo Bath House, also called "the Plunge," was constructed south of the Pavilion on El Paseo. It was 43,688 square feet and was boasted as being "the largest indoor saltwater pool in the world." About 2,000 people could bathe in one of its three heated saltwater pools. It was also equipped with Turkish baths and steam rooms. There were 1,354 dressing rooms with separate men's and women's changing rooms, of course. The women's dressing room was located on the second floor and had three women attendants on hand constantly. This early-20th-century locker room had hot and cold tub baths, showers, and massage and hairdressing parlors.

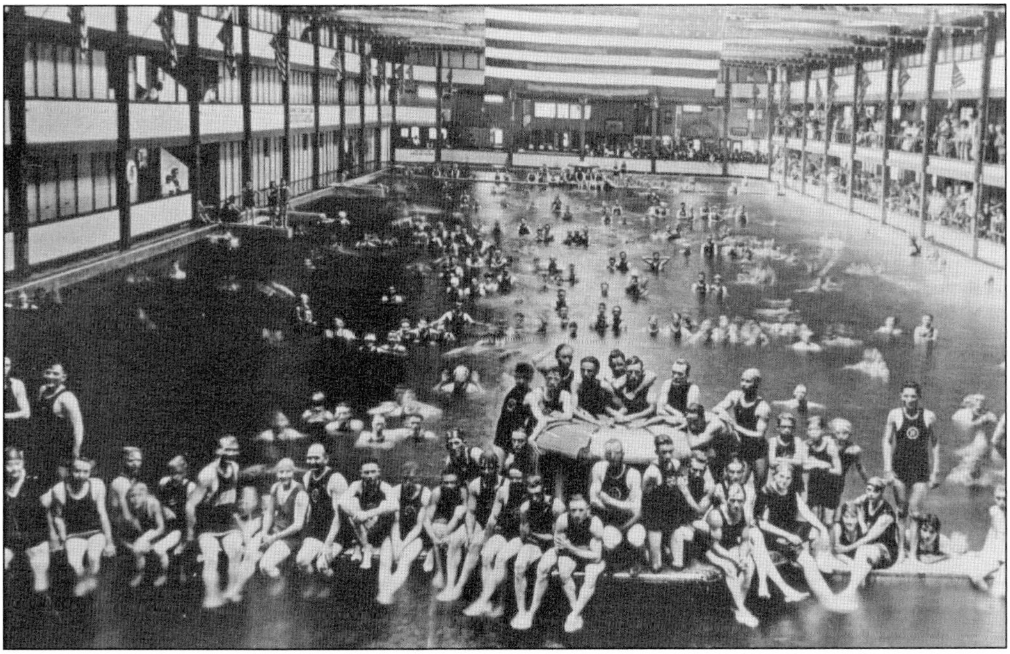

The grand opening of the Bath House was on July 1, 1909. Events included a professional water polo game between the Venice Beach and Redondo Beach teams, an underwater swimming race between famed Redondo Beach lifeguards Harvy Crum and L. Lawson, and a "fire dive" by a professional diver from the Venice Plunge. George Freeth, a lifeguard at the Bath House, showed off his diving skills for the crowd, and there were several amateur swimming competitions. Two million gallons of water were poured into the Plunge daily. The drying and laundry department included 8,000 swimsuits available for rent and 20,000 towels. Seven thousand incandescent lamps illuminated the interior and several thousand more lined the exterior. Seen in these photographs are interior shots of the Plunge. The image above shows hundreds of people in the pool and three levels of bathing rooms. Below is a photograph of the giant slide that was clearly very popular with visitors.

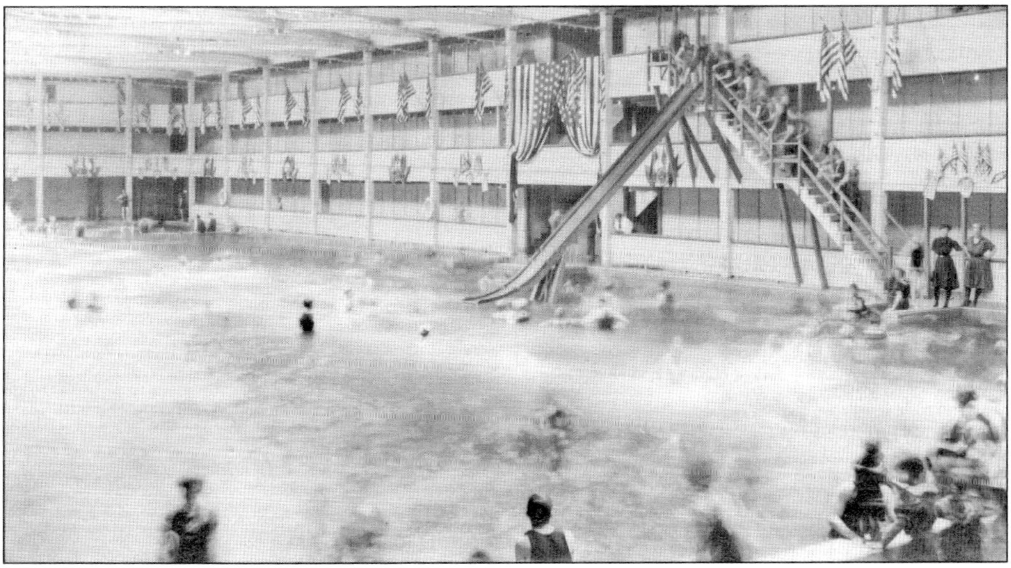

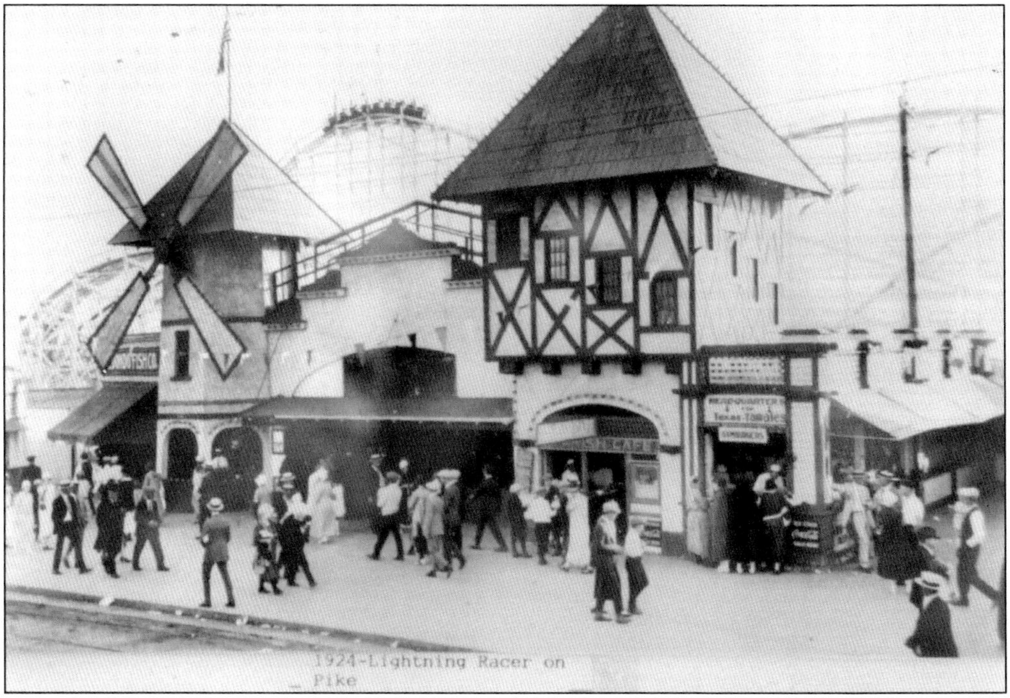

This is the entrance to the amusement zone located north of the Pavilion. The grand entrance has a German influence to it, but the half-timbering and windmill are purely decorative. At one time, this was an entrance to a ride called the Rapids. In the background of this 1920s photograph, the Giant Dipper roller coaster can be seen.

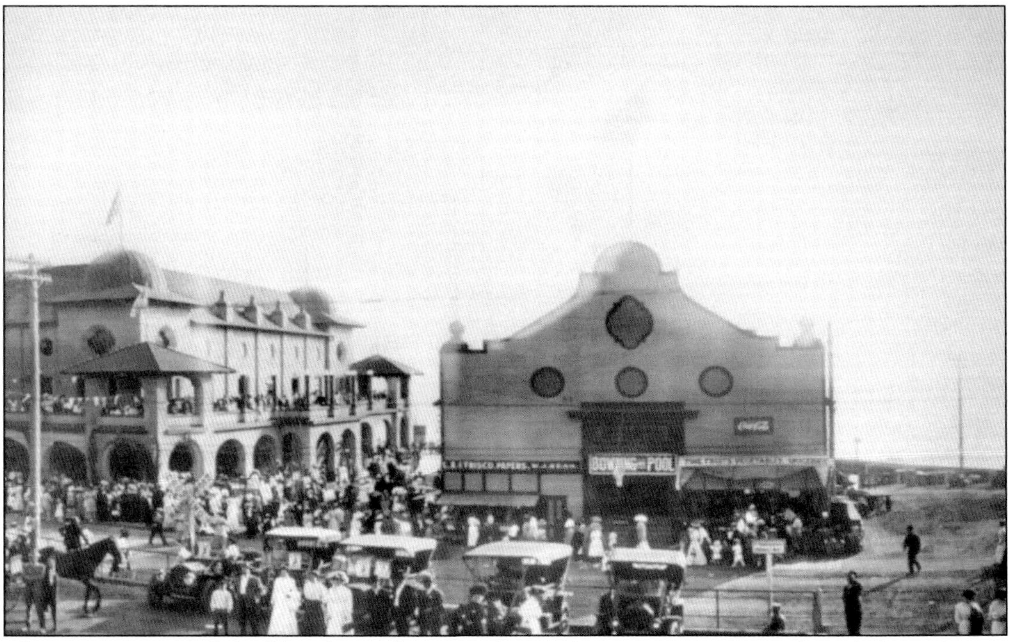

This is a Mission-style building located next to the Pavilion some time around 1905. According to the sign on the building, it housed a bowling alley and most likely pool tables. The Mission-style parapet and quatrefoil medallion are decorative elements complimented by the Pavilion.

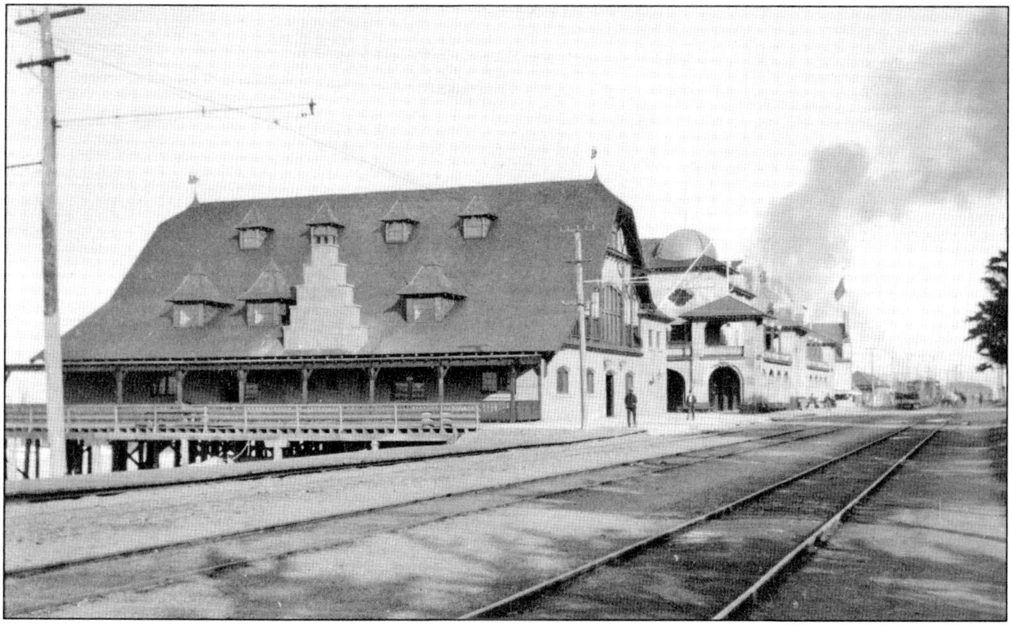

This photograph of the casino was taken around 1915. The building, like the entrance to the Rapids, is of German influence with its large sloping roof, pronounced dormer windows, stepped chimney, and the half-timbering underneath clipped gable ends. A *Los Angeles Times* article refers to the building as a café and restaurant in the early 1900s. Later images show that it was at one time the Capitol Theatre, then the Art Theater.

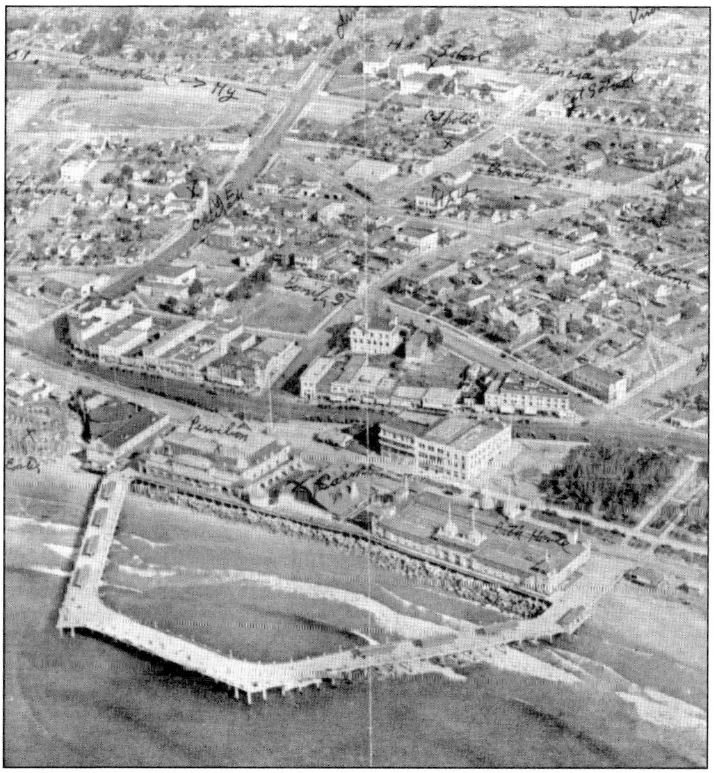

Because the pleasure pier was inherently used for leisure and entertainment, it only made sense that the shape of the pier should accommodate the amusing ambiance. The Endless Pier was built in 1915 in the shape of a horseshoe. It offered restaurants, cigar and candy shops, bandstands, ballrooms, dance halls, and bowling alleys. This is a photograph of the pier around 1916. The Bath House and the Pavilion can both be seen.

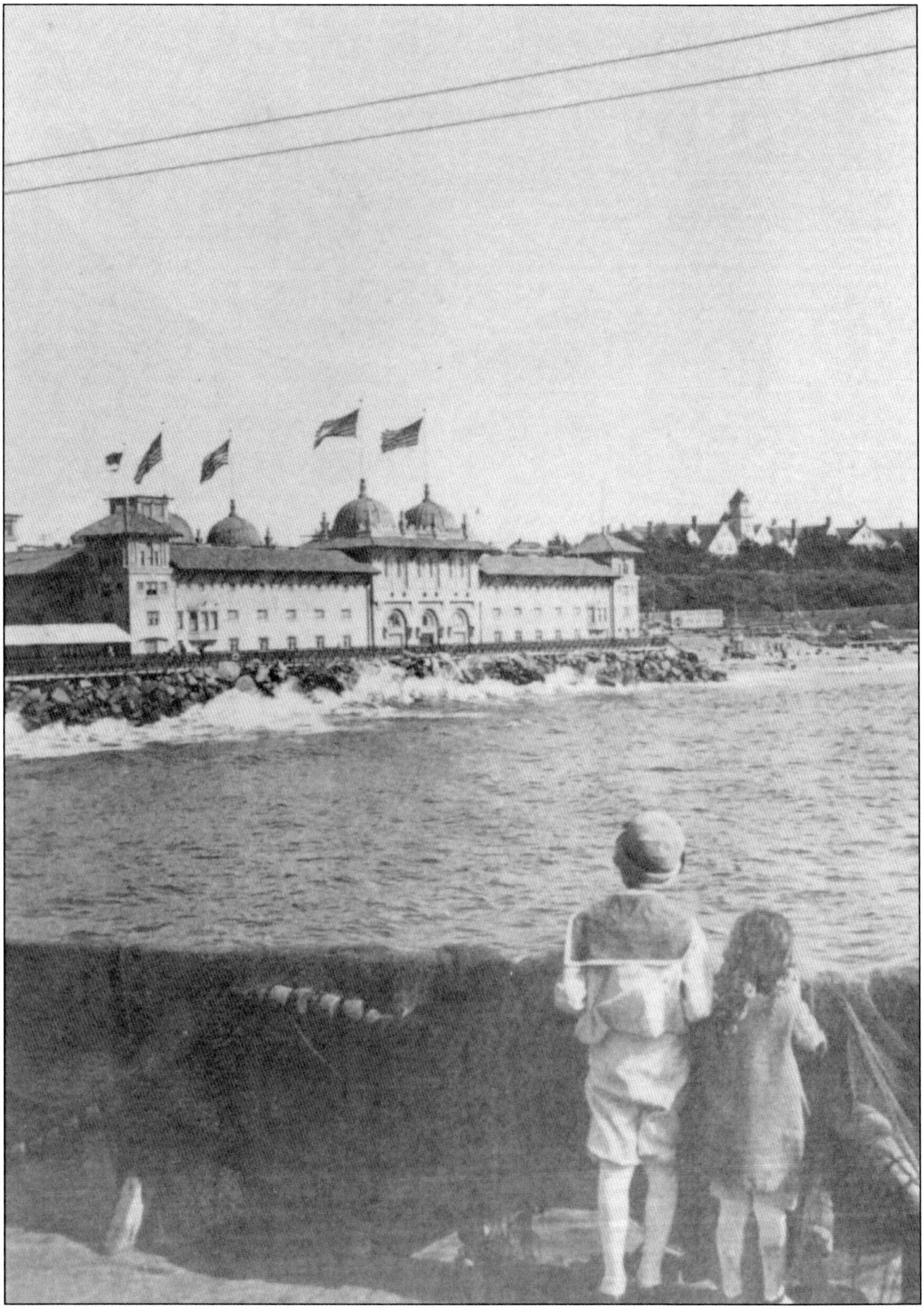

The styles of the seaside architecture in Redondo Beach in the early 20th century reflect a high-style architecture that evokes moods of pleasure, wonder, relaxation, and escape from everyday life. The history of the pleasure pier in Redondo Beach transformed the recreation of the beach resort in the early 20th century.

Four

THE DECLINE OF THE PLEASURE PIER

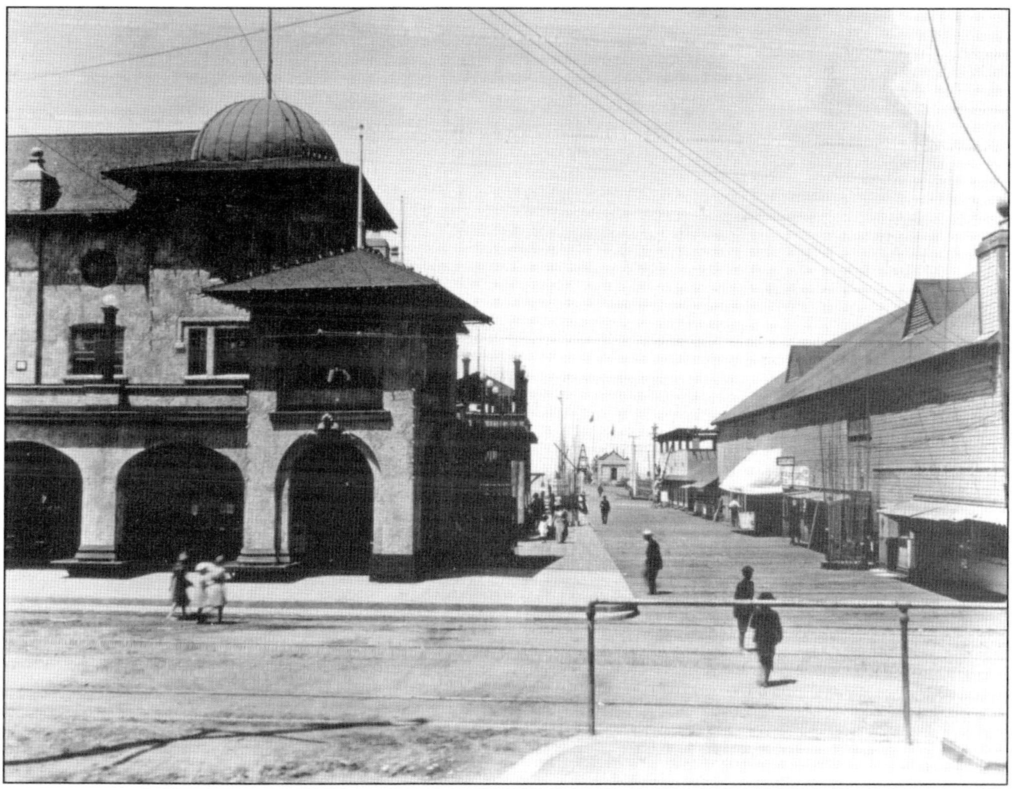

The heyday of amusements, entertainment centers, and venues lasted along El Paseo well into the 1950s. Around the late 1930s, gambling became prominent along El Paseo, and the entertainment strip eventually transformed into a less than family-friendly environment. As such, business began to decline. However, in the 1940s, the Pavilion and other venues were still in operation. The Pavilion housed an amusement center, a roller-skating rink, and arcades, but El Paseo was no longer the lively entertainment strip bustling with tourists that it once was.

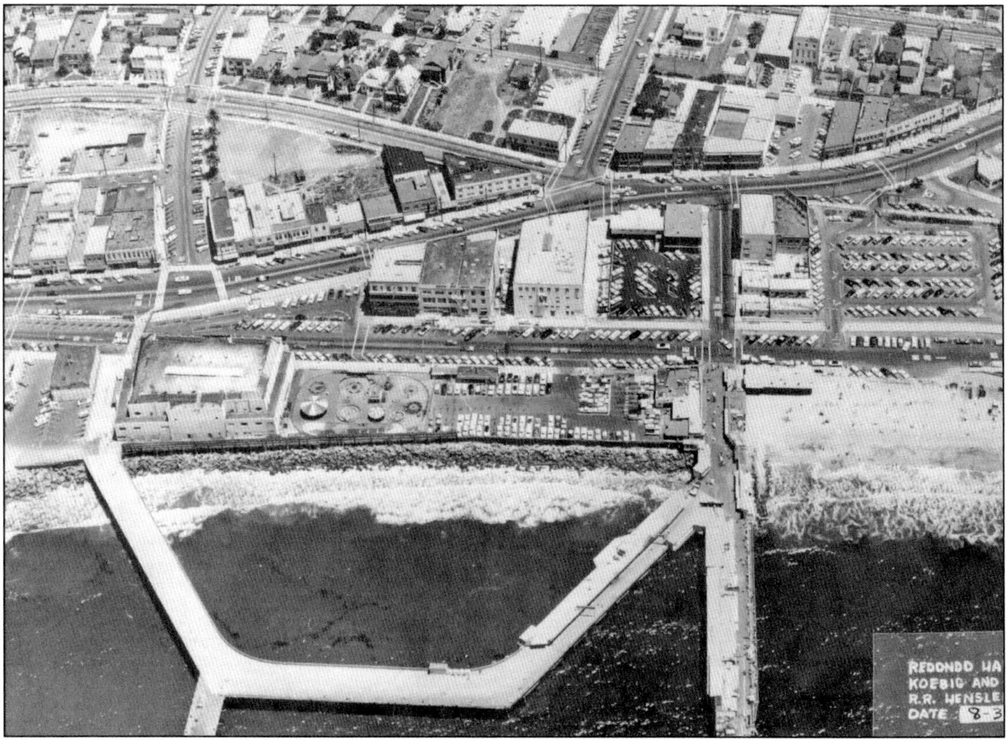

The advent of World War II and especially the bombing of Pearl Harbor caused the United States to place military presence along its western coast. From San Diego to Washington State, the West Coast piers and ports were transformed into military zones. While Redondo Beach did not have a huge military presence in comparison to Long Beach and San Diego, tourism was certainly affected. Blackouts and military drills were practiced in the city, and it was no longer the bustling tourist attraction of roller coasters and entertainments. Above is an aerial view of Municipal Pier in the 1940s. The Plunge was demolished in 1941, and the Pavilion has been refaced but remains. Below, the Pavilion is being restored with a new facade. The Plunge has already been torn down.

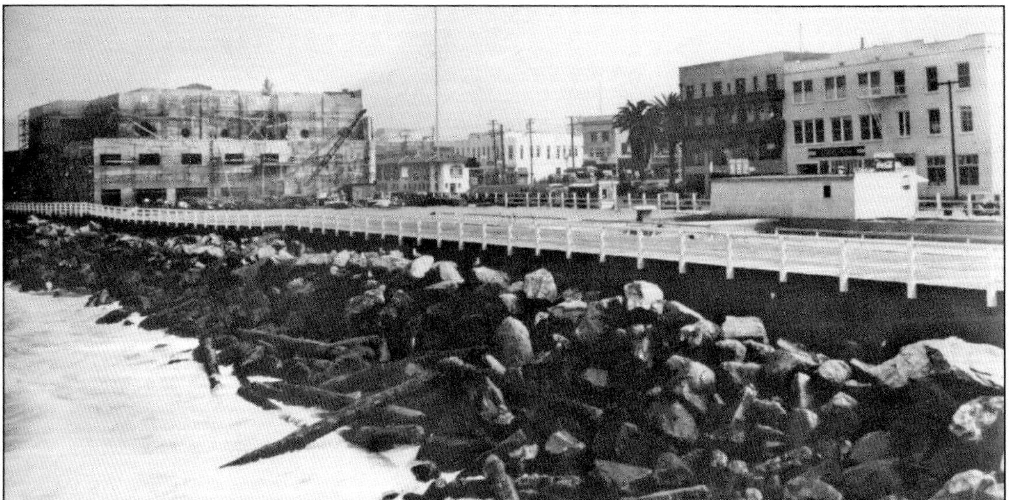

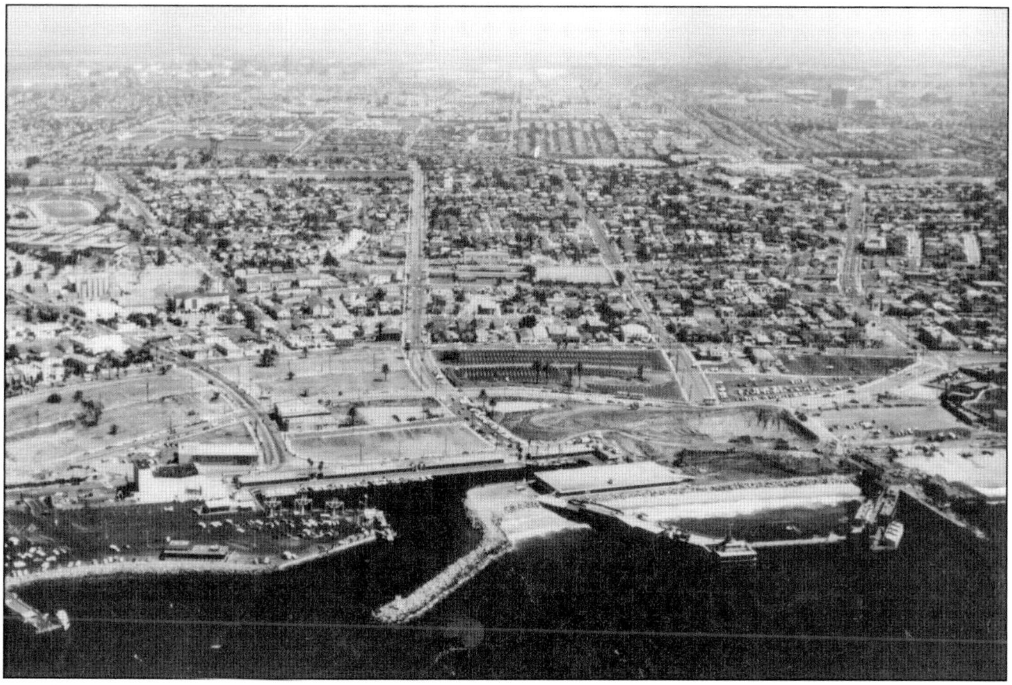

Parallel to the decline of the beach resort towns, the Pacific Electric Red Cars railway was replaced by major highways, known as freeways, and the automobile. By the 1950s, most of the beach resort towns were taken off the trolley route. Today, Redondo Beach can only be accessed by car. Nonetheless, limited parking and traffic have changed the number of people that could once access the beach via multiple trainloads per day. The photograph above illustrates the redevelopment that occurred in the 1950s. El Paseo's midway is gone, and the area was redeveloped for a parking lot. Below is a view of the parking lot by the pier in the 1950s.

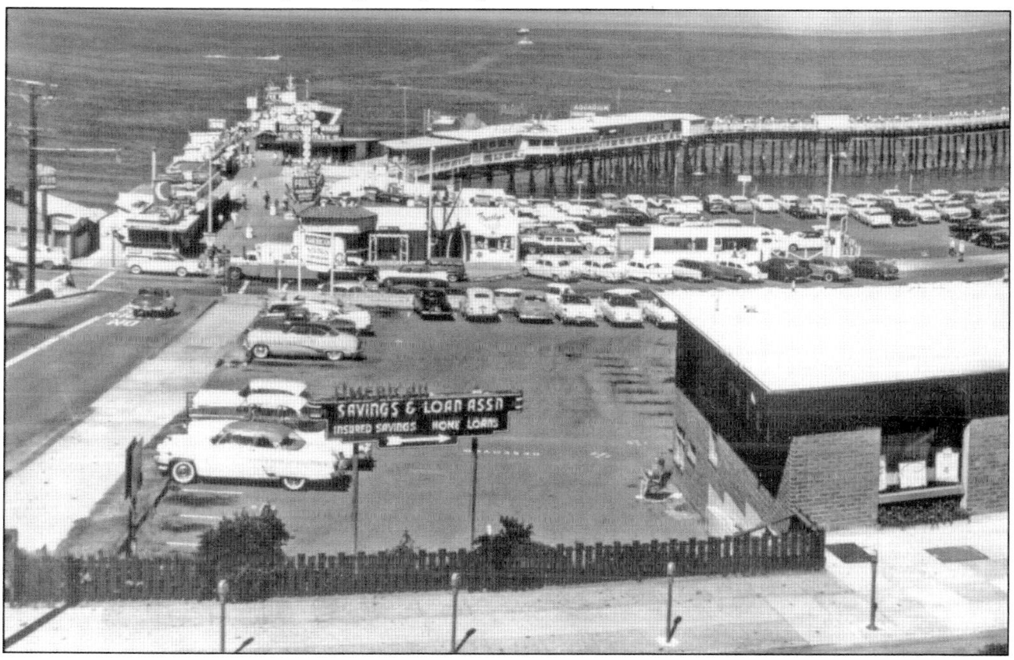

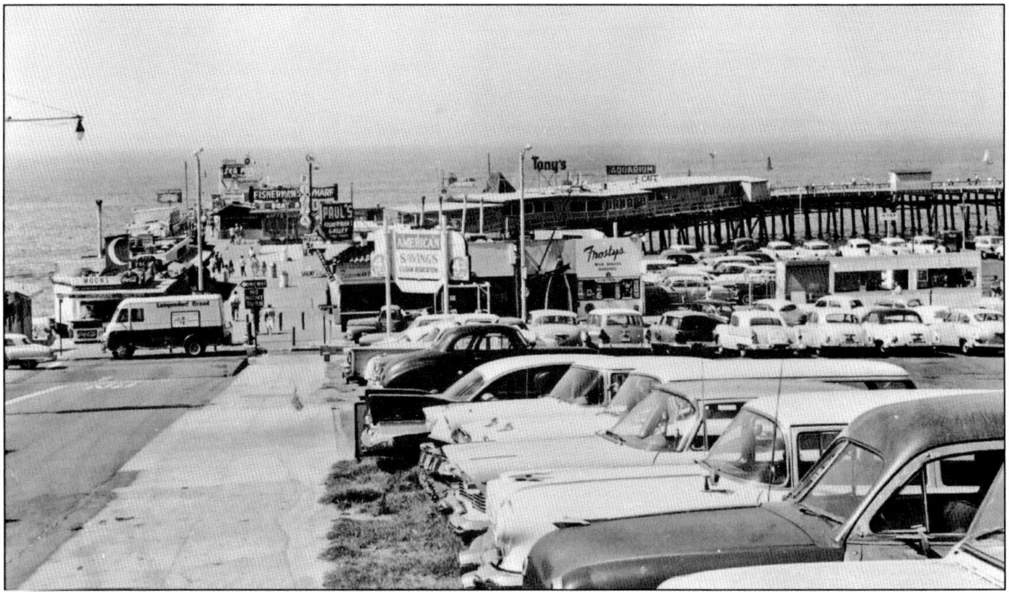

This is how the Redondo Pier appeared in the 1950s. Note the amount of cars as well as the apparent rise in fishing activity since the days of the amusement zone. (Courtesy of the Redondo Beach Historical Society.)

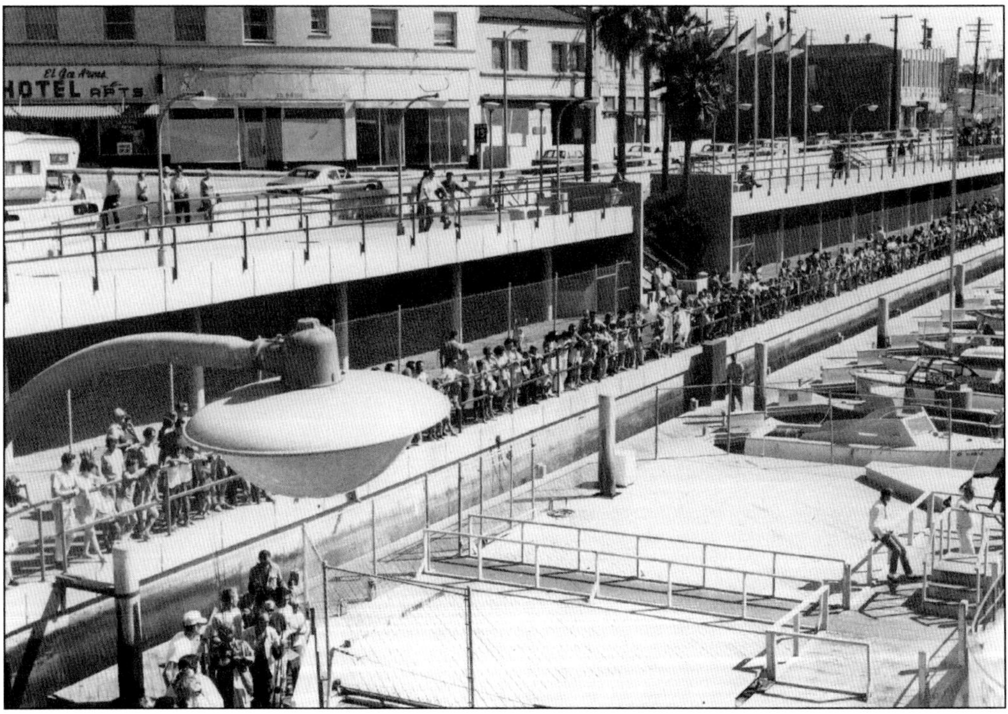

This photograph highlights the redevelopment of El Paseo in the 1950s and 1960s. The small rectangular marina that one can see today was dredged in the 1950s and 1960s. There are now fishing boats and docked yachts where the amusement area once stood. Note that the buildings are still standing on what was once Pacific Avenue. Later, these buildings were replaced by high-rise condominiums.

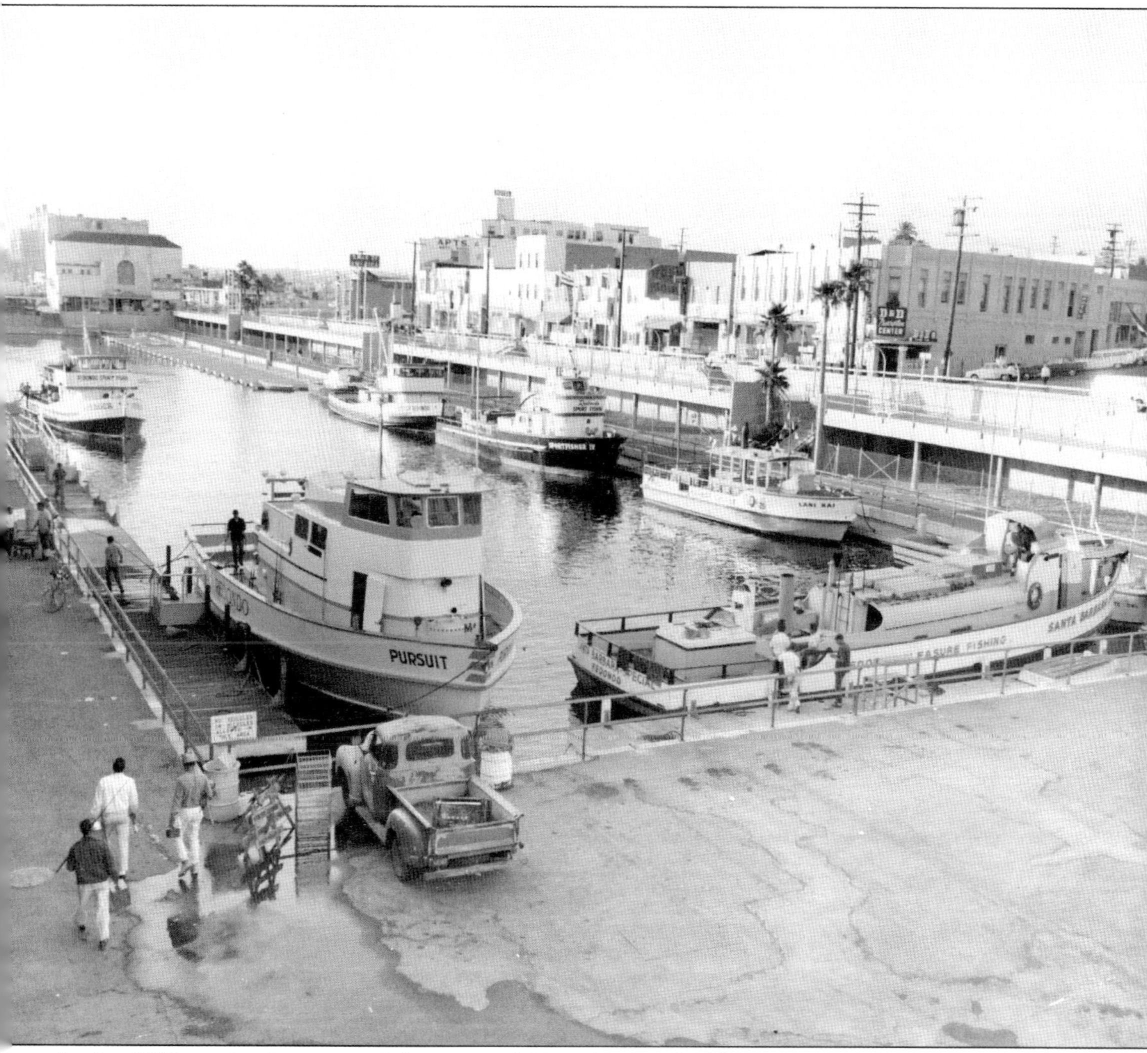

By the 1950s, redevelopment of the pier and King Harbor had begun. Above is the area where the amusement zone along El Paseo once stood.

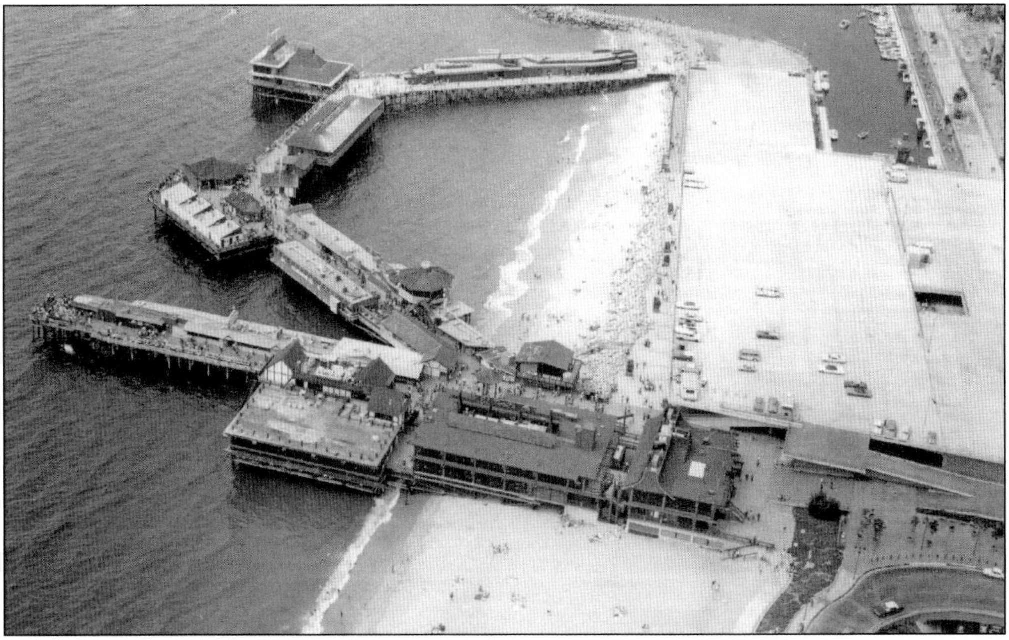

In 1955, Walt Disney proposed that the Magic Kingdom be established in the city of Anaheim in Orange County. The amusement park had been 20 years in the making for Disney and was to be a big economic boon for Anaheim and the county. To everyone's surprise, the Magic Kingdom was more popular than originally anticipated. Anaheim's land value skyrocketed and business boomed immediately. Tourism shifted from Southern California beaches to Disneyland. Redondo Beach's amusement park setting was already gone, so focus was shifted to redevelopment and a new harbor. Here are two aerial photographs of Municipal Pier in the early 1970s.

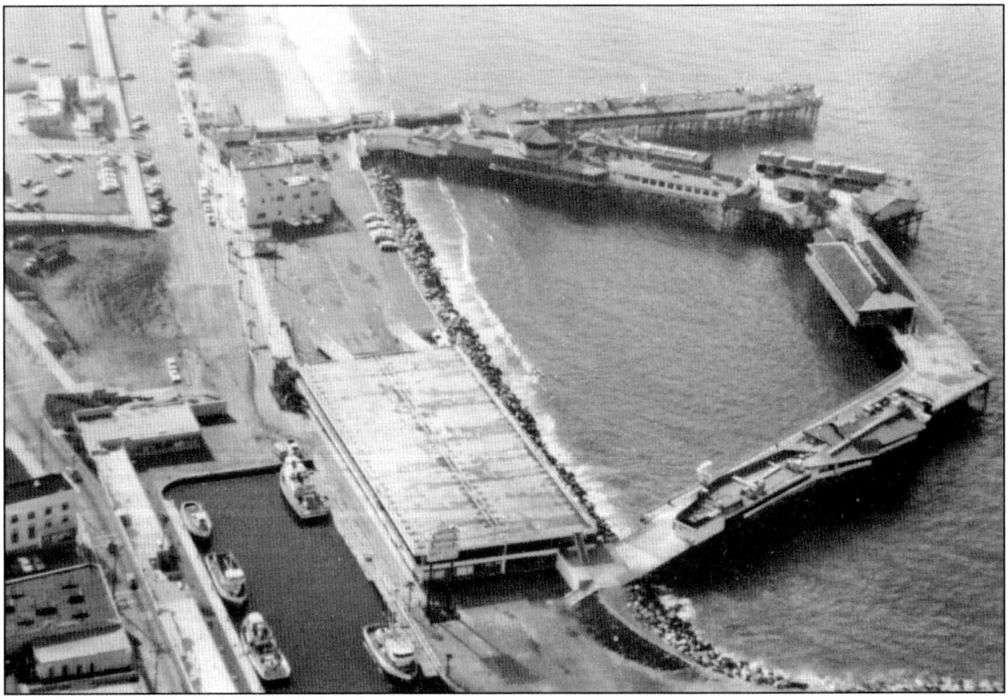

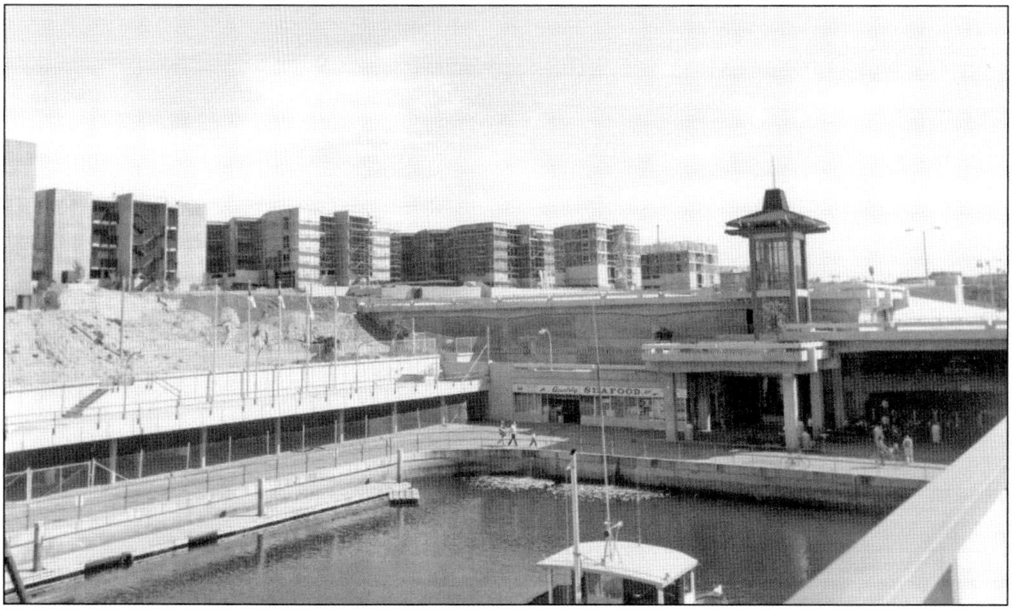

The 1960s and 1970s was a prime time for redevelopment in Redondo Beach. During this period, El Paseo was no longer an entertainment zone but was transformed into condominiums and housing. Unfortunately, sometimes redevelopment means tearing down the old to make way for the new. Much of the old beach resort–era architecture has been replaced with newer condominiums and modern housing. However, if anything, this redevelopment shows the growth of community in Redondo Beach.

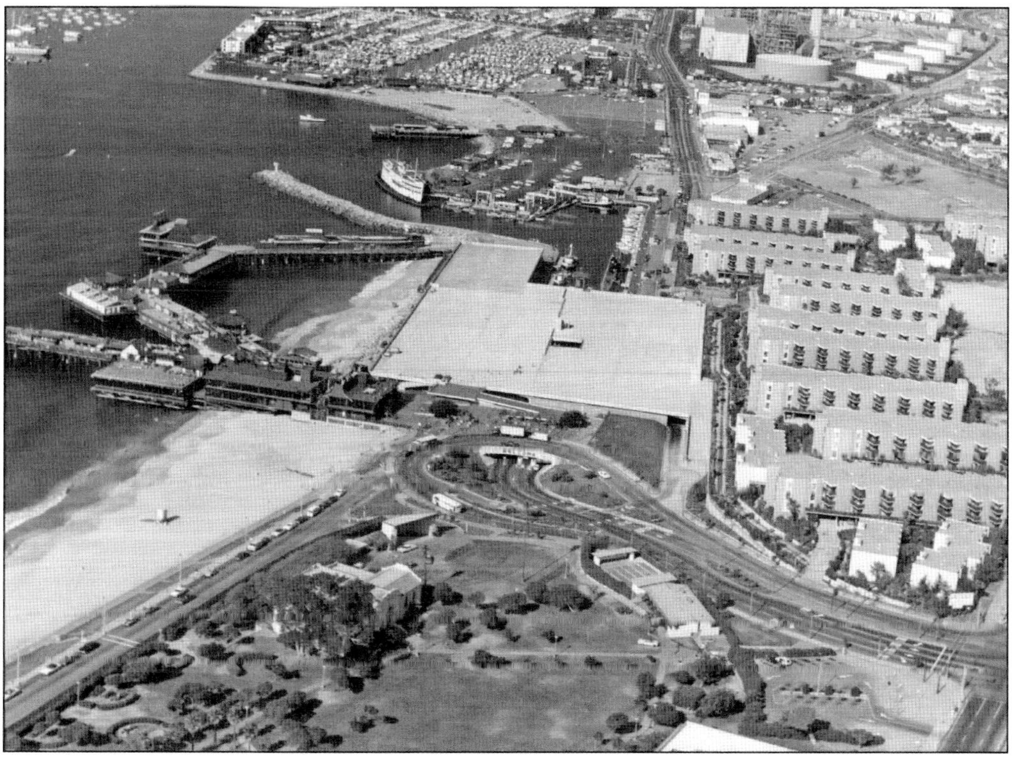

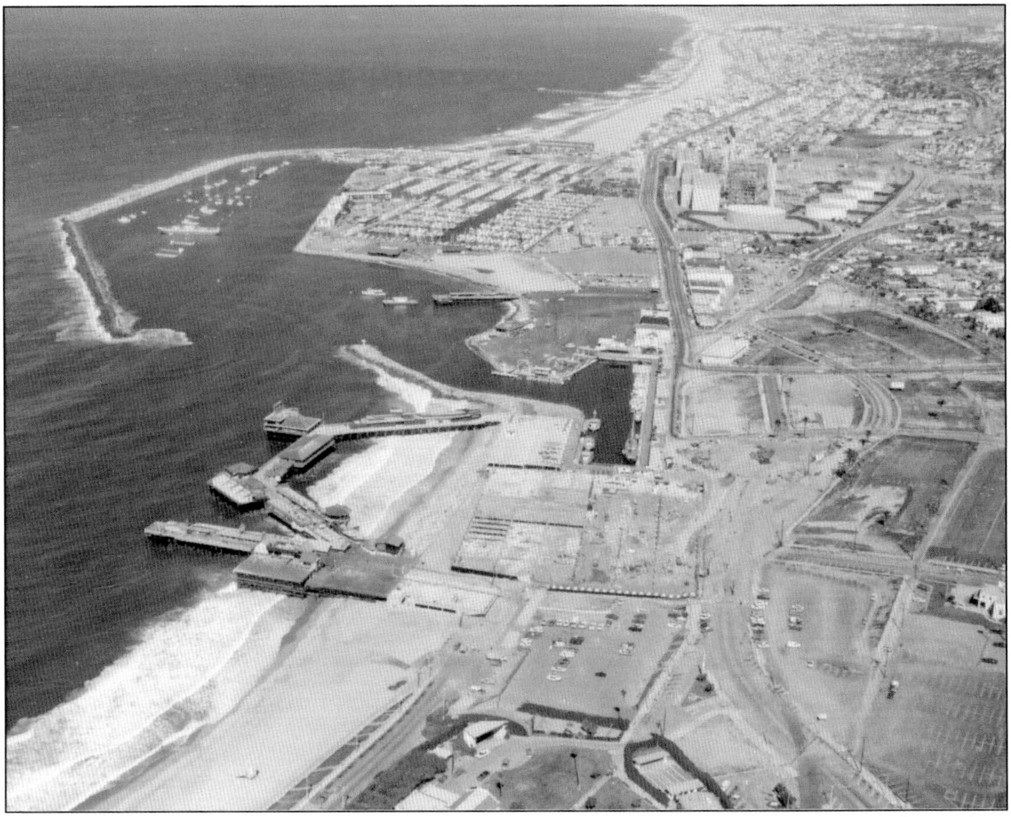

As early as 1935, Redondo Beach started planning for a recreational harbor port for yachts, canoes, kayaks, and other small craft. In 1938, a breakwater was constructed to help dissipate the torrential effects of storms that hit the south bay. Above is an aerial shot of new King Harbor in the 1970s.

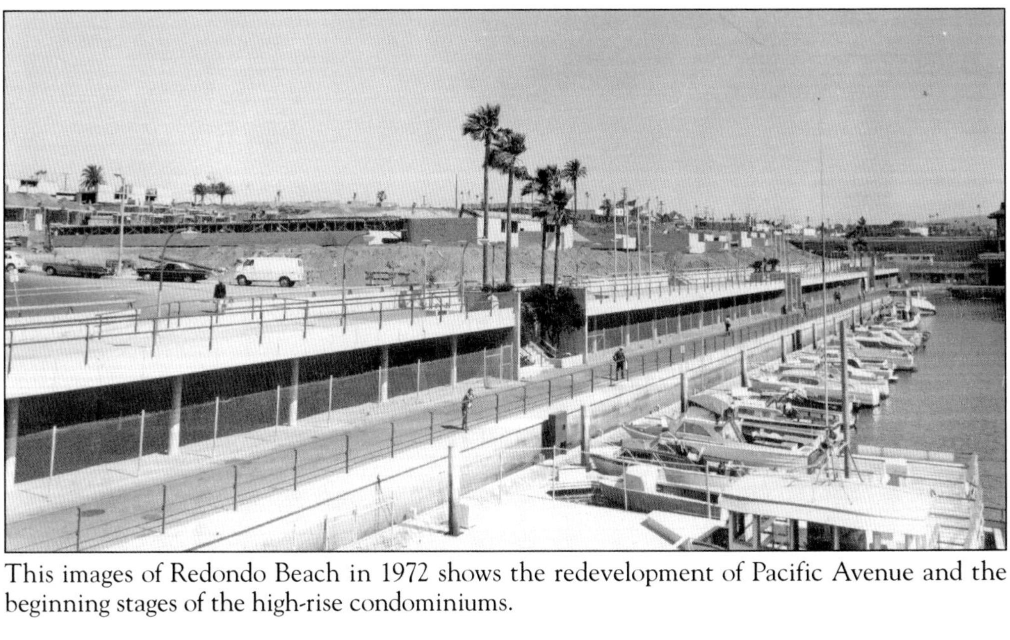

This images of Redondo Beach in 1972 shows the redevelopment of Pacific Avenue and the beginning stages of the high-rise condominiums.

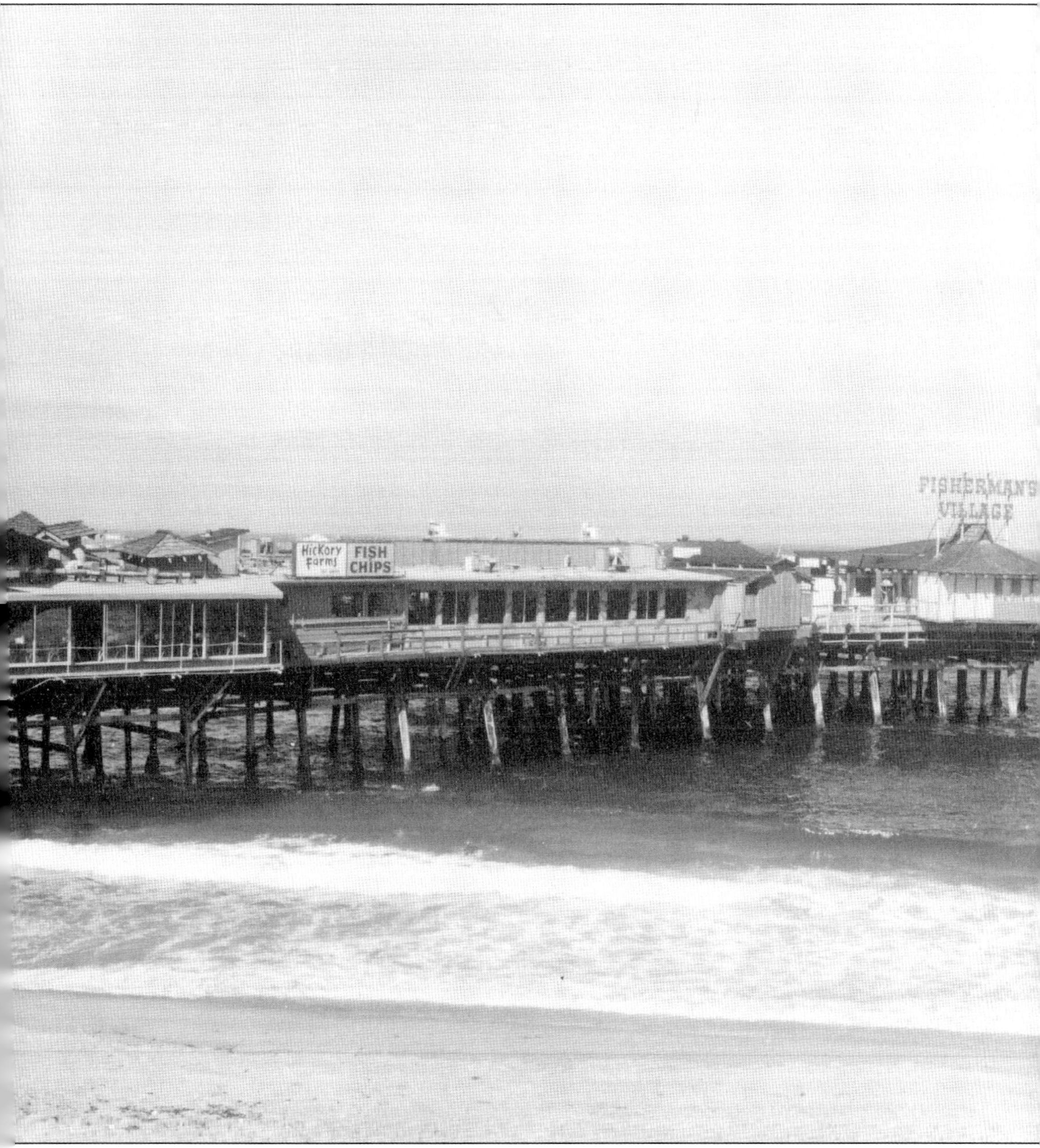

Above is the Redondo Pier in the 1960s. It was and still remains a popular place for fishing. In 1964, city officials of Redondo Beach collectively named the pier Fisherman's Wharf. Today, it is called Municipal Pier.

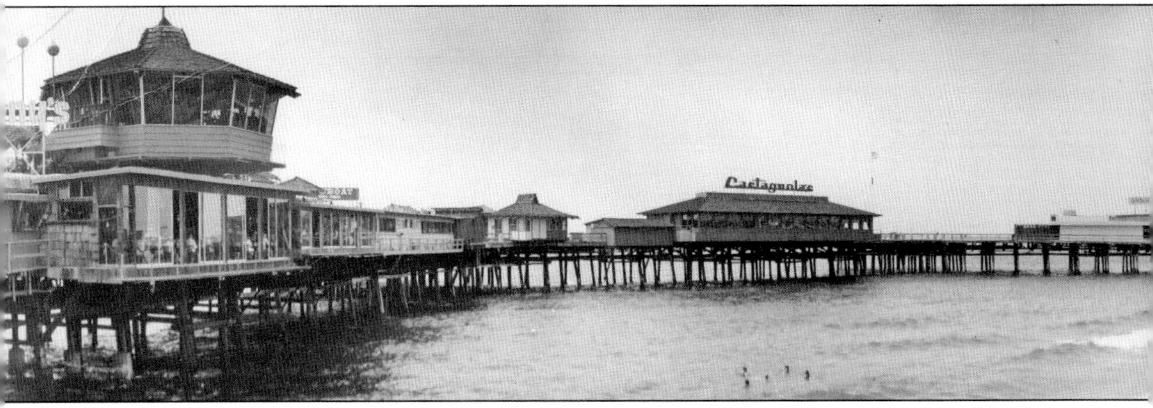

Municipal Pier has undergone several changes since it was originally constructed in 1915. Storms and fires have destroyed the pier over the decades, but it has always been rebuilt. The area, first developed in the early 20th century, has seen redevelopment. While the pleasure pier is no longer an amusement zone as it once was, it is still the main attraction for tourists in Redondo Beach. This photograph of Municipal Pier is from the 1980s.

Five

The Redondo Beach Municipal Pier Today

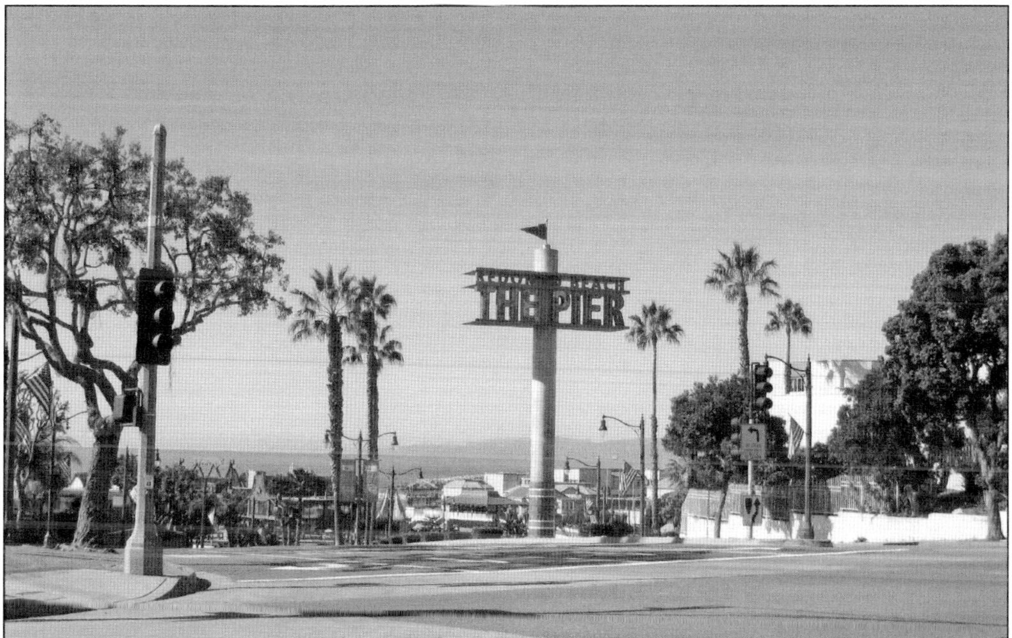

The Redondo Beach Municipal Pier still attracts thousands of tourists every year. Some of the restaurants and venues may have changed over the years, but the atmosphere of relaxation and fun remains.

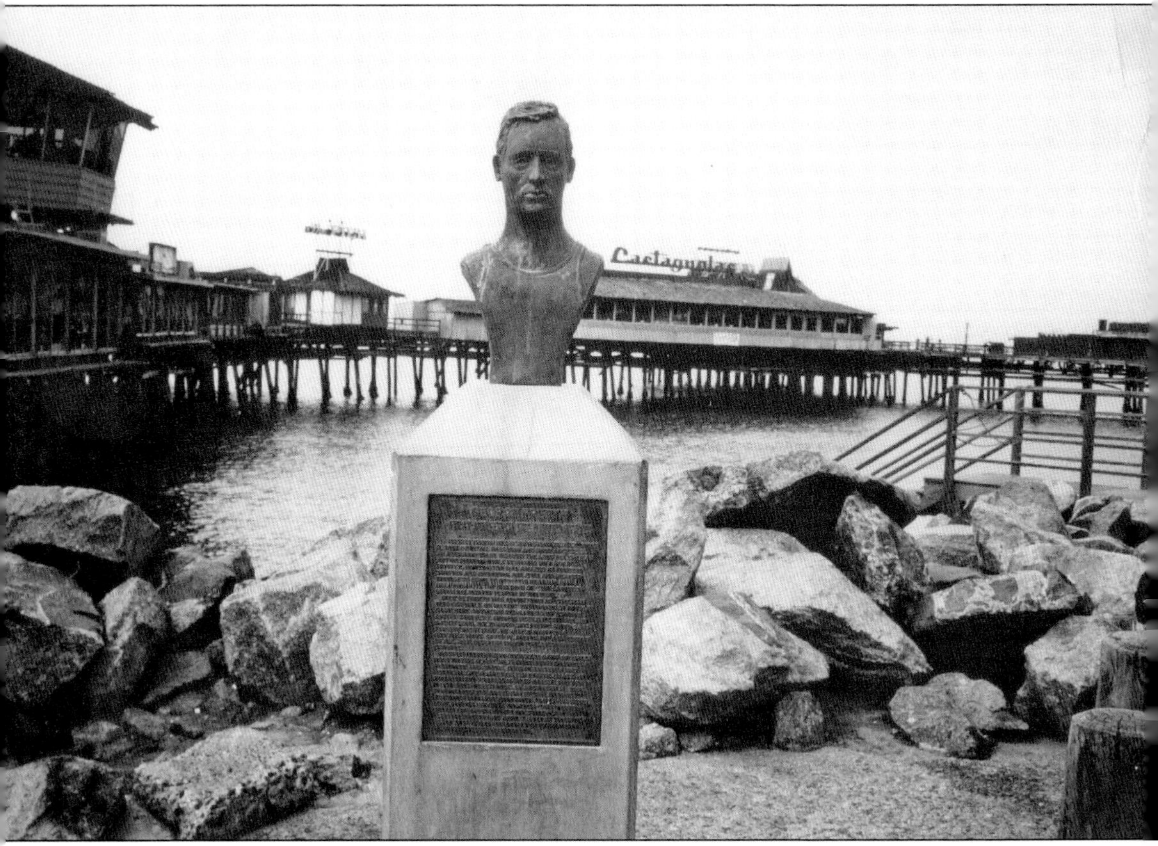

In 1976, a bronze bust of George Freeth, the legendary surfer and lifeguard of Redondo Beach, was placed on a concrete pillar next to Municipal Pier in honor of his service. In August 2008, the bust, which was 100 pounds of bronze metal, was stolen and presumably melted for scrap. The Friends of Freeth, which included the Redondo Beach Historical Society and neighboring businesses and residents, raised $22,000 to have the bust reproduced. The unveiling of the new bust took place in November 2010. Above is a picture of the old Freeth statue.

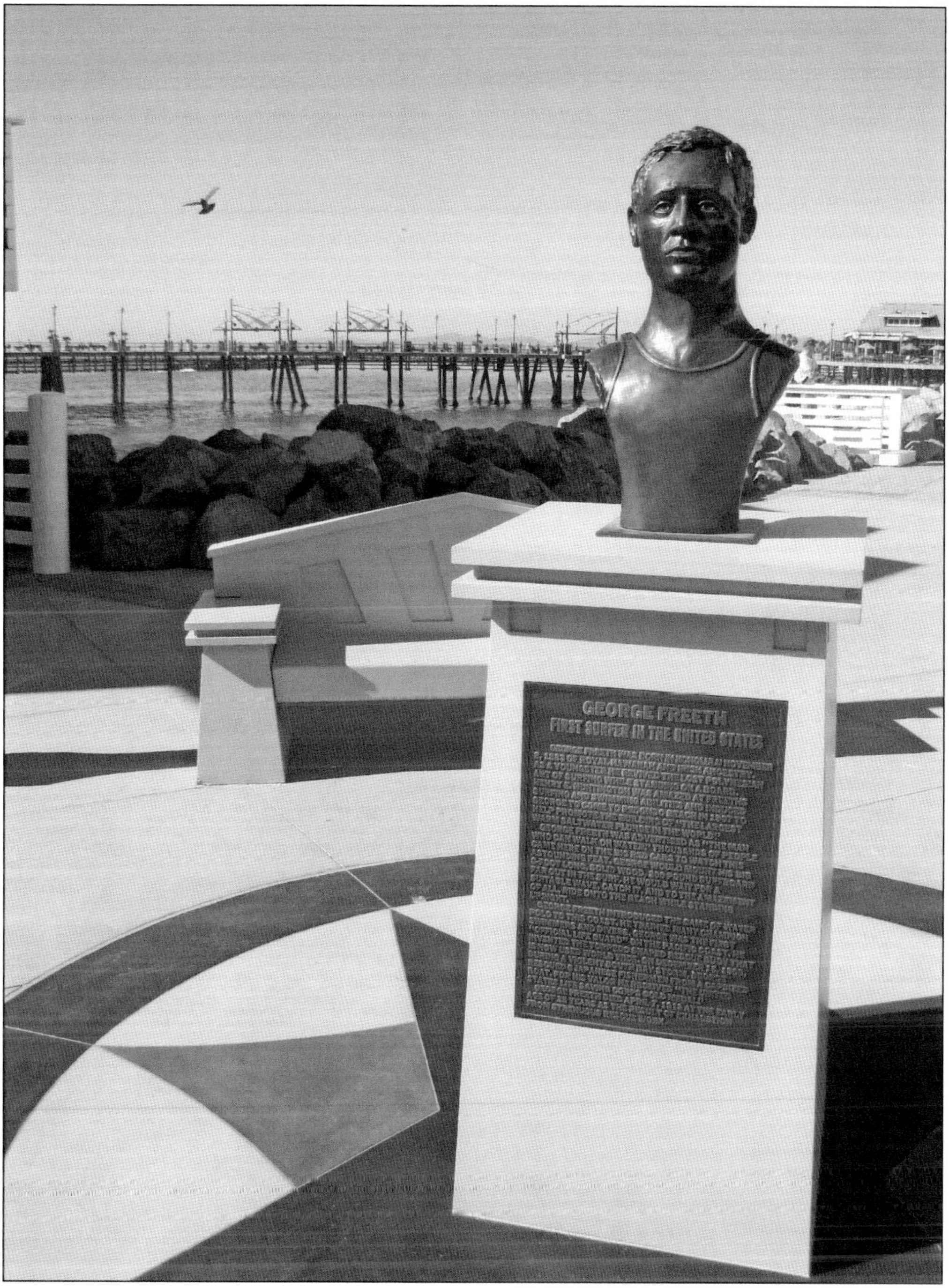

Pictured above is the new George Freeth bust in November 2010. (Photograph by Author.)

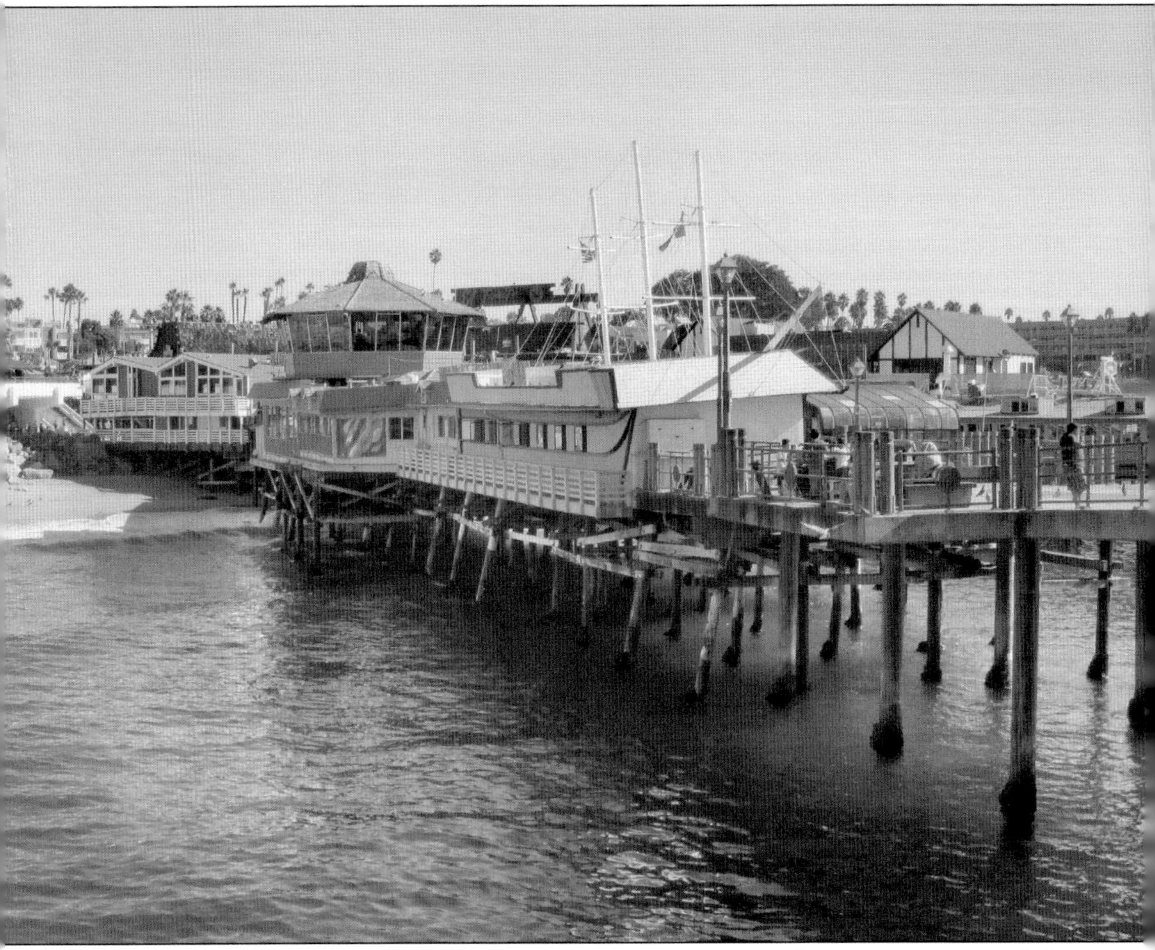

Today, Municipal Pier is still a lively venue for entertainment and restaurants. Pictured above is the south leg of the pier. The original shaded settees and benches have been replaced with restaurants, souvenir boutiques, and music venues. (Photograph by Author.)

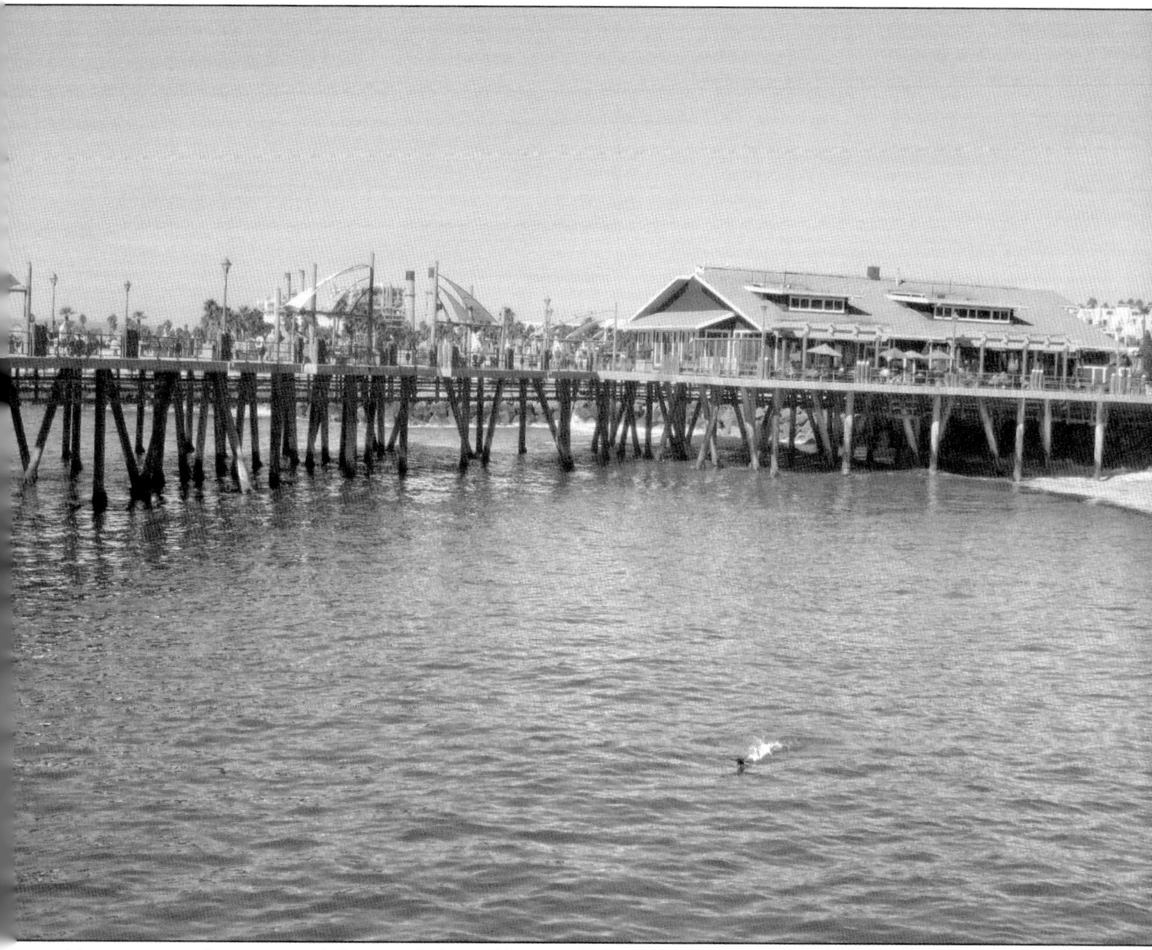

Above is another view of the south leg of Municipal Pier. This area of the pier is adorned with picnic benches for patrons to sit on and enjoy the water. At any given time, fishermen can be seen casting their reels for the catch of the day. At the end of the pier is a seafood restaurant. (Photograph by Author.)

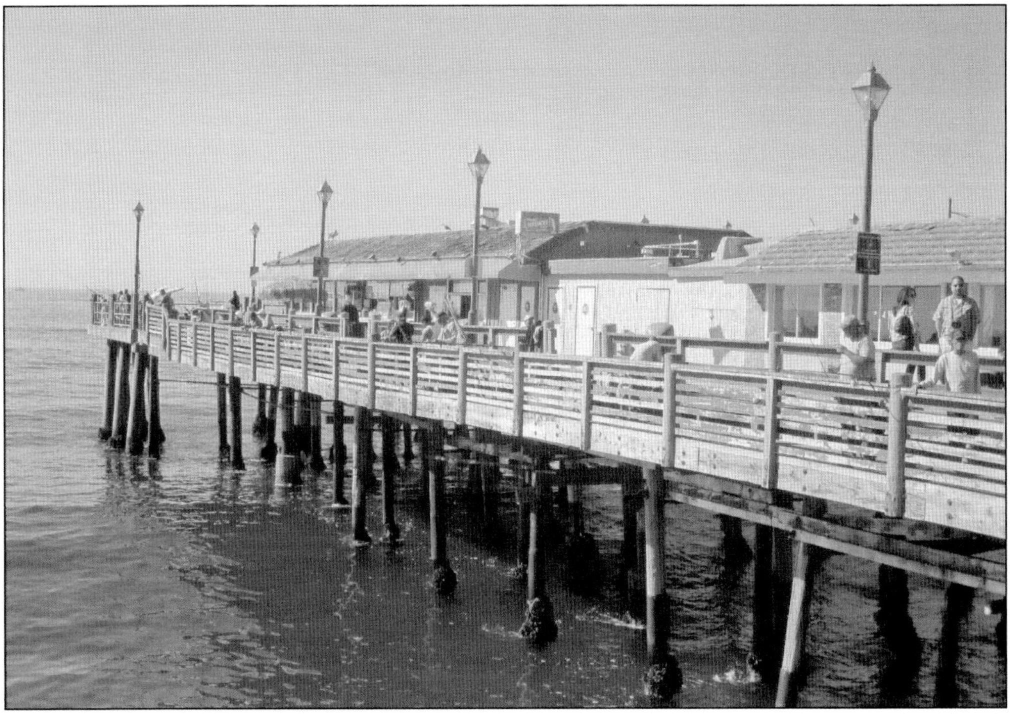

Pictured above are remnants of the Monstad Pier today. This popular place to fish is lined with seafood eateries. (Photograph by Author.)

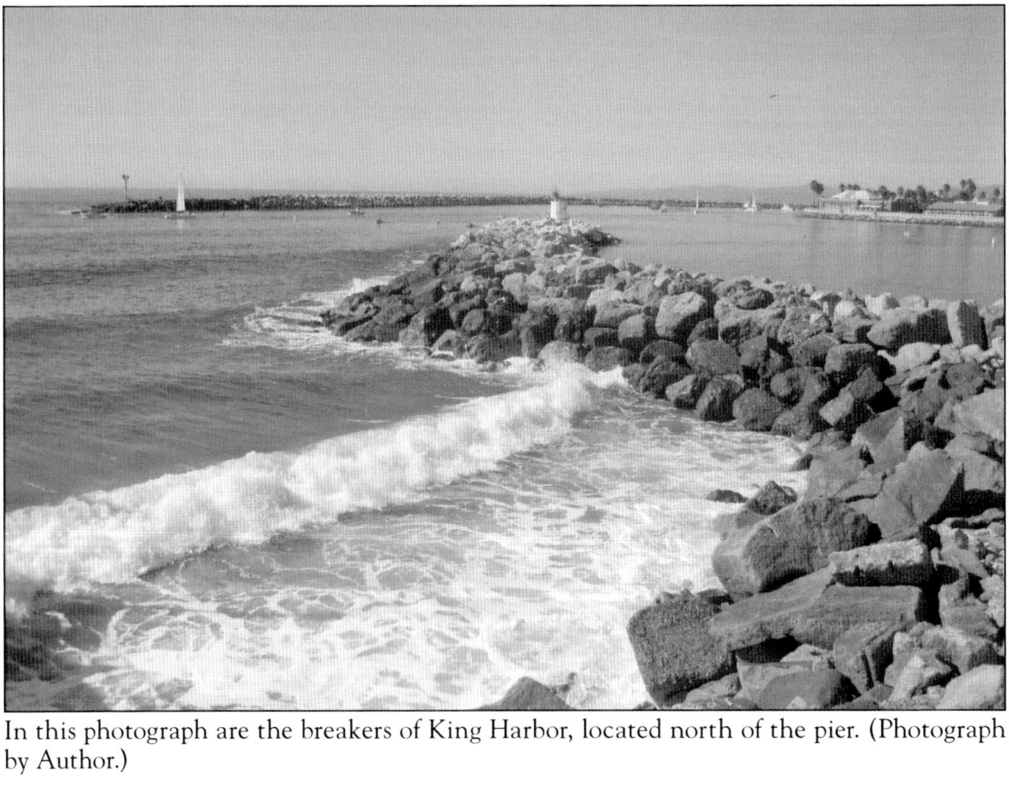

In this photograph are the breakers of King Harbor, located north of the pier. (Photograph by Author.)

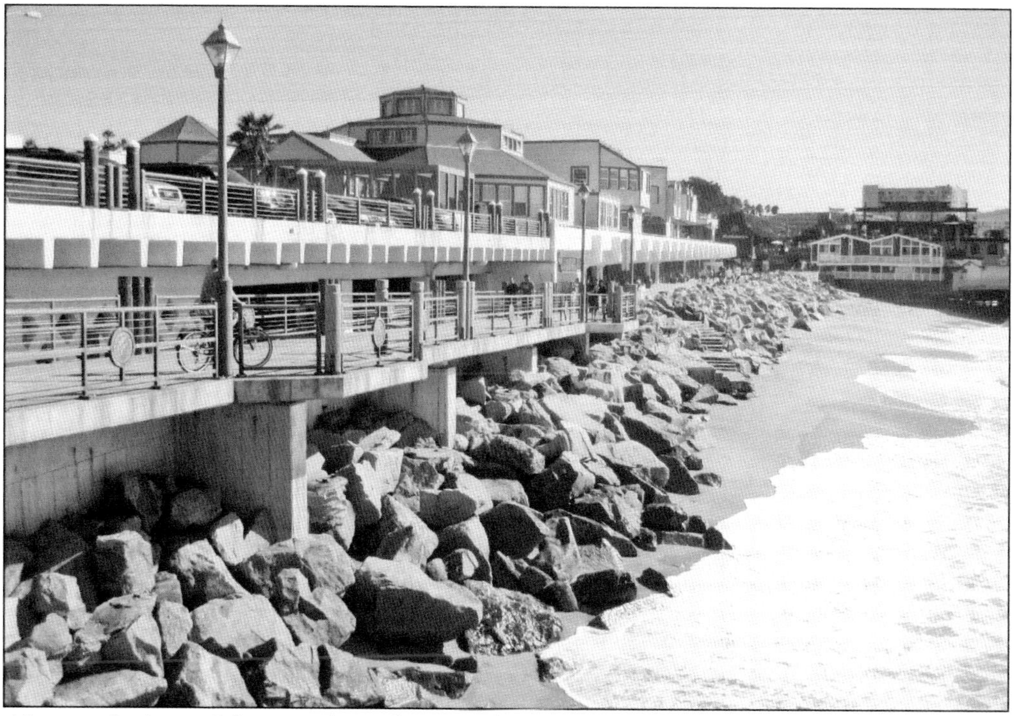

Above is the base of the pier where El Paseo, the Pavilion, and the Plunge once stood. Today, it is a parking facility and office park. (Photograph by Author.)

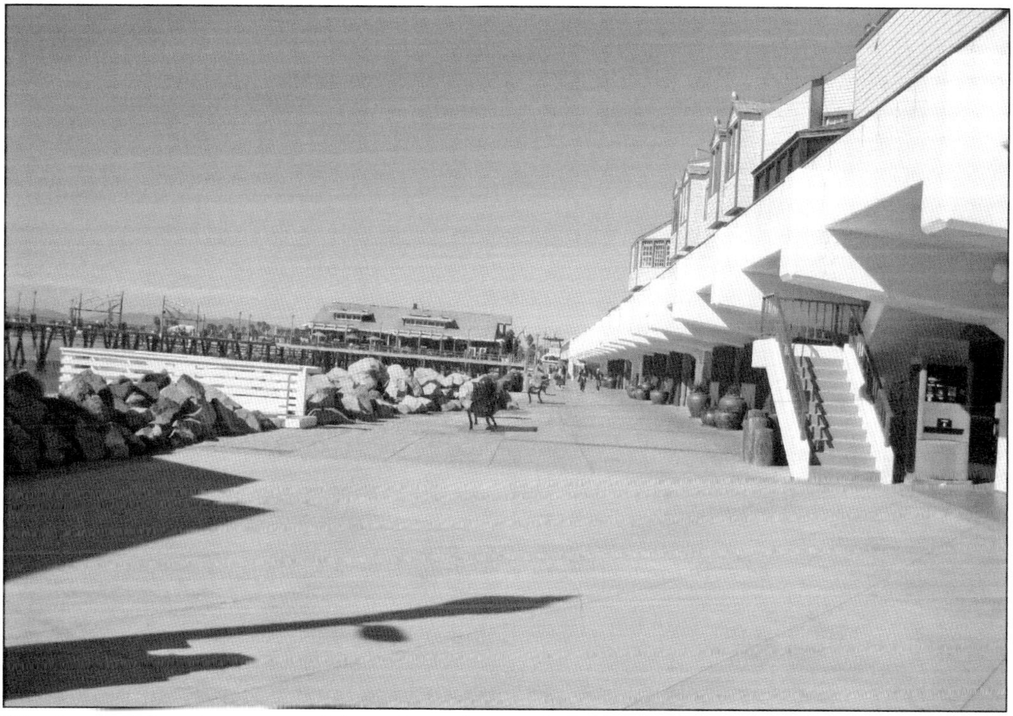

The base of the pier is viewed here from the south entrance of Municipal Pier. (Photograph by Author.)

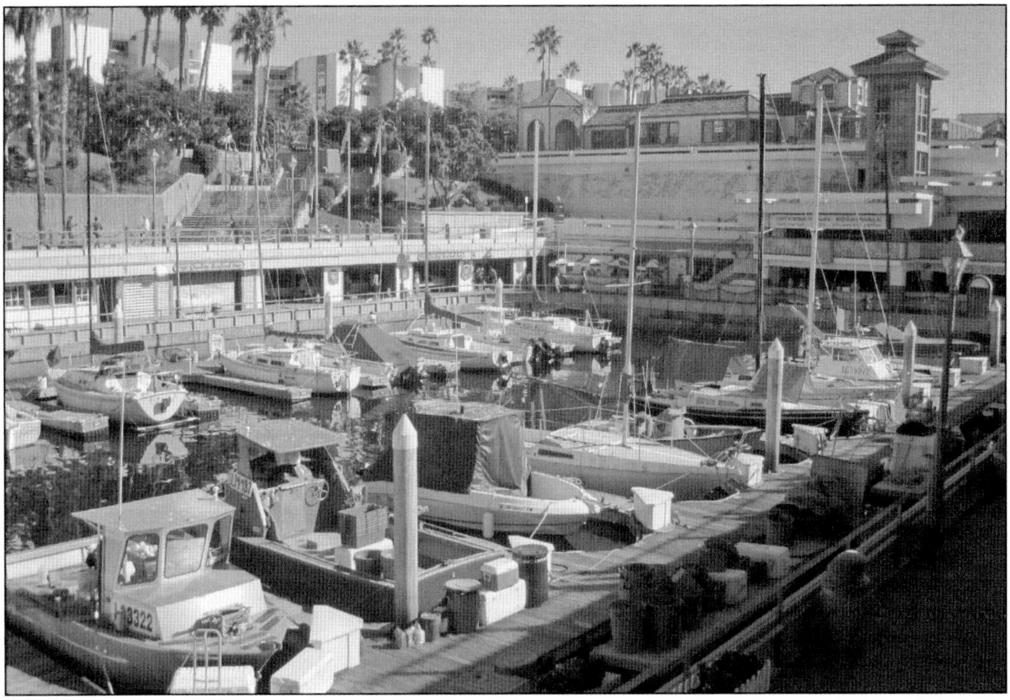

Shown above is the International Boardwalk and a marina for small yachts and fishing boats. In the photograph below, condominiums are located on what used to be Pacific Avenue. (Photographs by Author.)

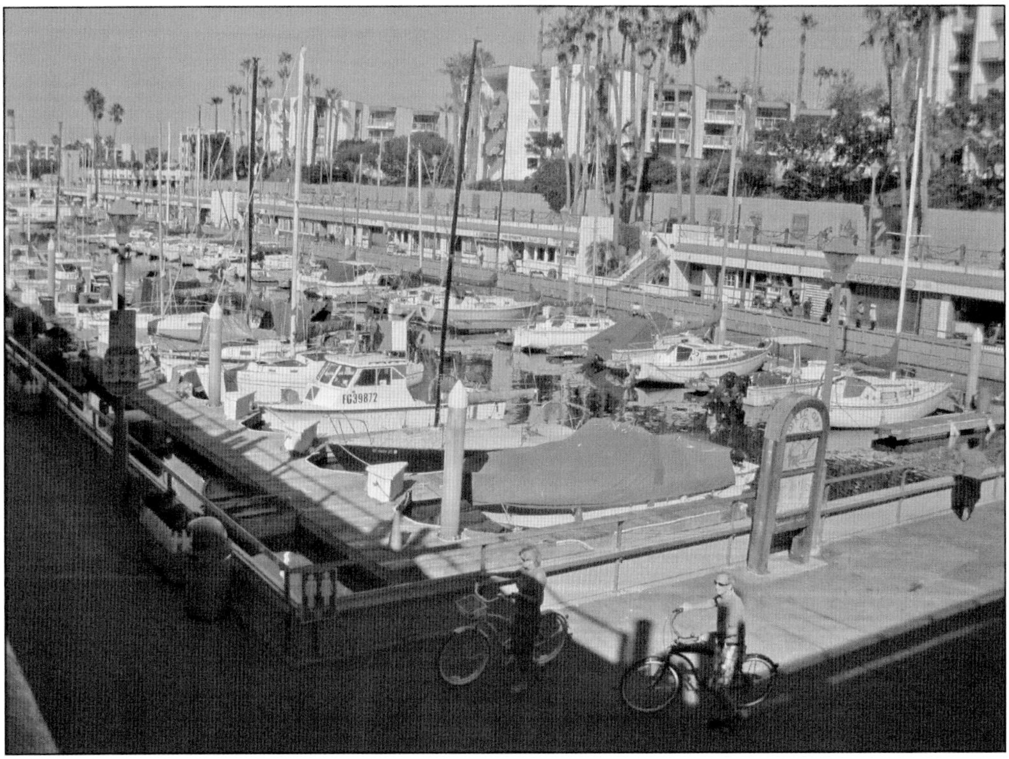

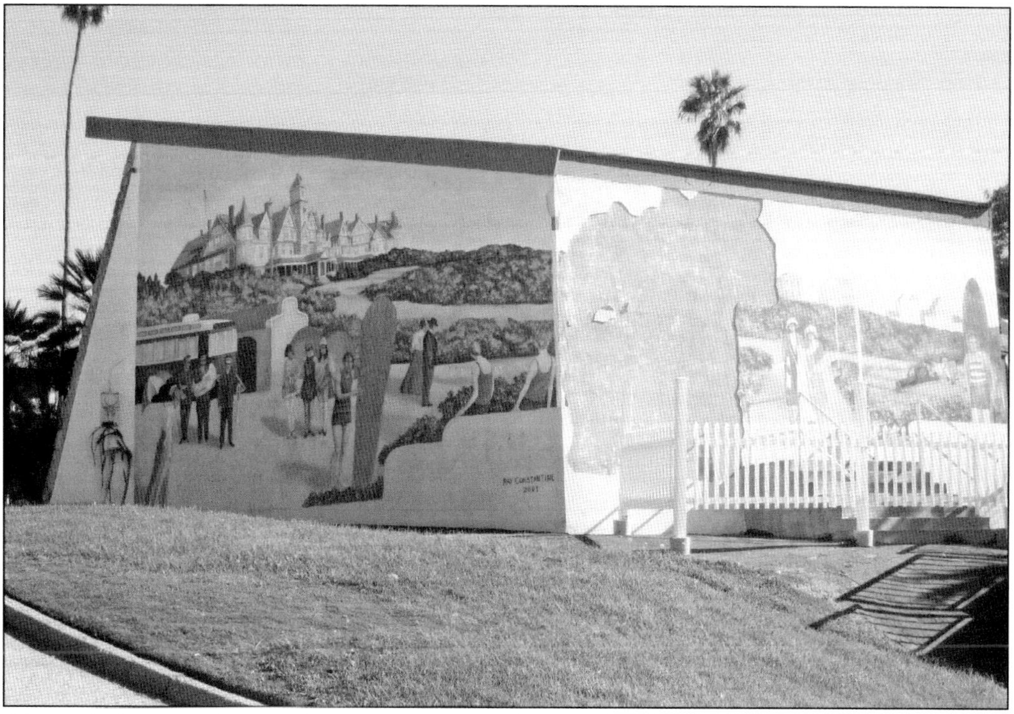

Although Hotel Redondo is long gone, the mural painted on the back of this bandstand reflects how the area looked in the early 20th century. (Photographs by Author.)

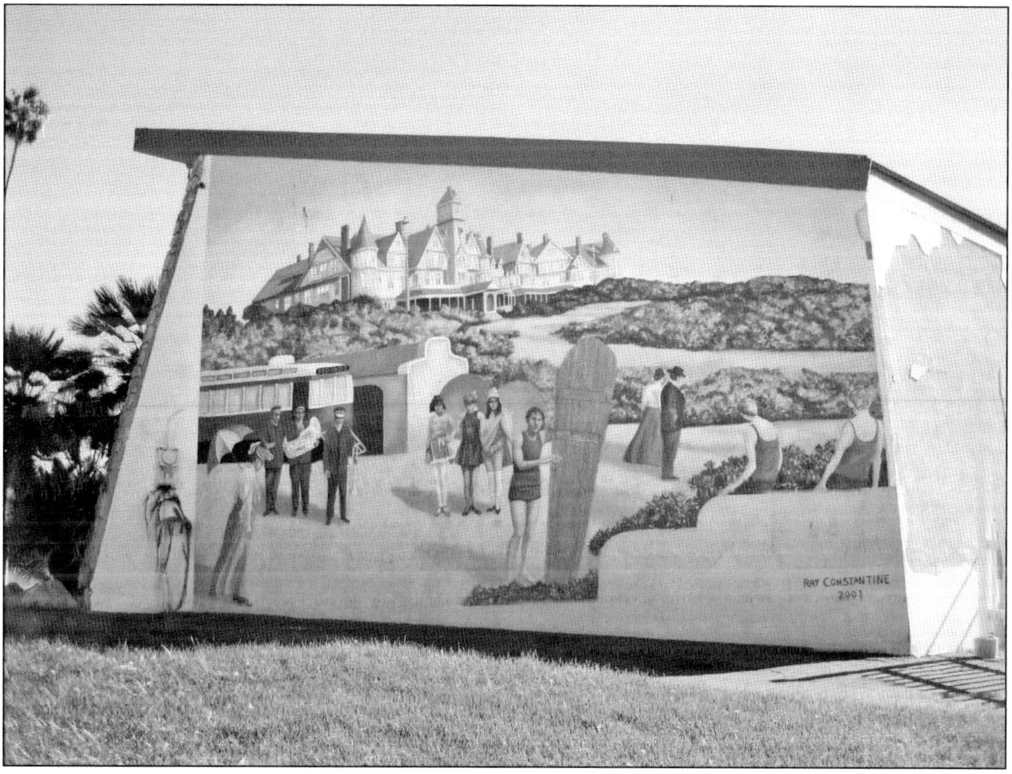

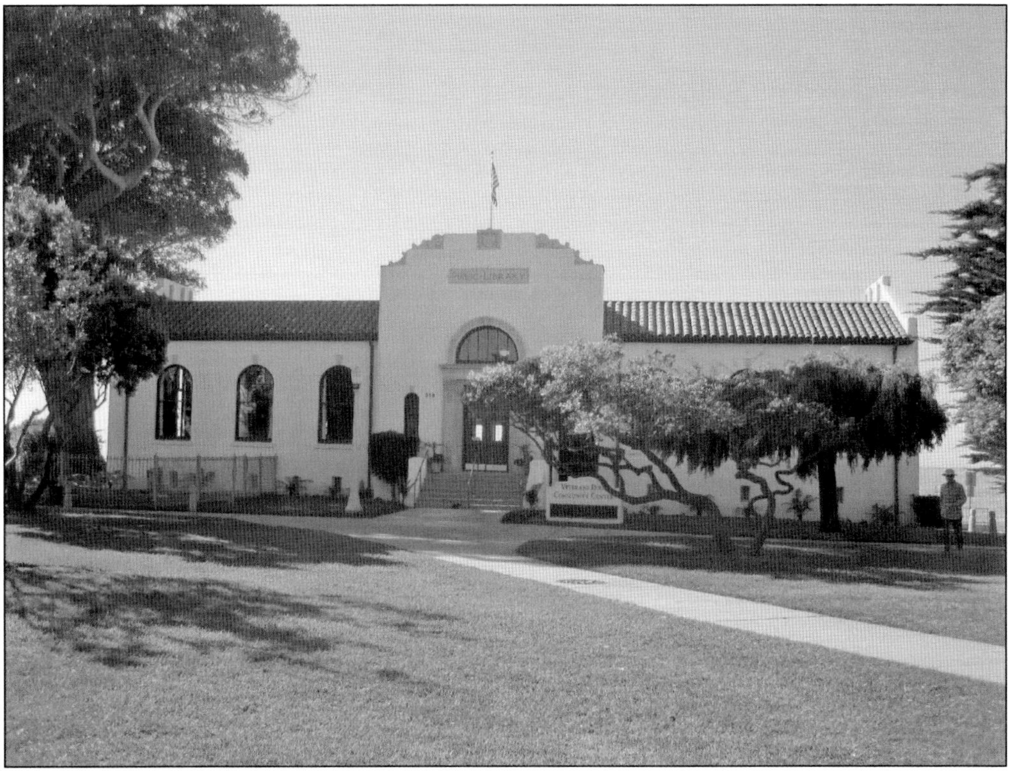

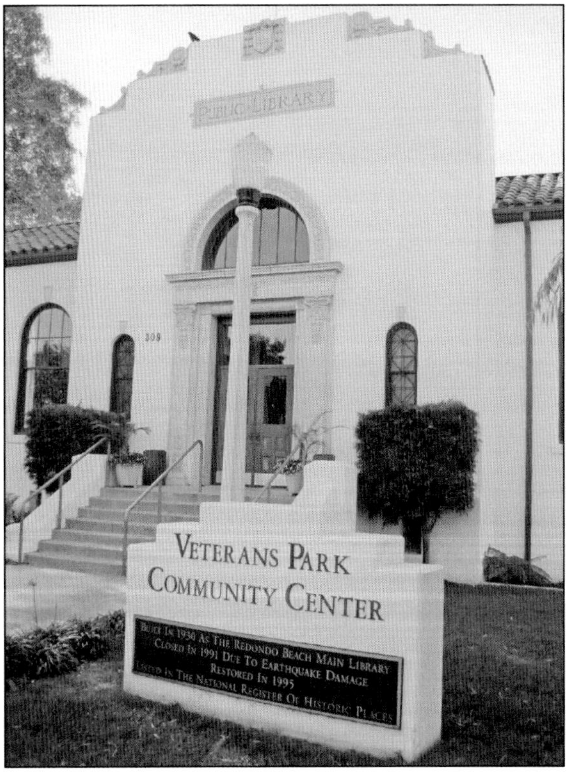

Hotel Redondo was replaced by the Redondo Beach Library in 1930. The building is now the Veterans Park Community Center in Veterans Park and is listed on the National Register of Historic Places. (Photographs by Author.)

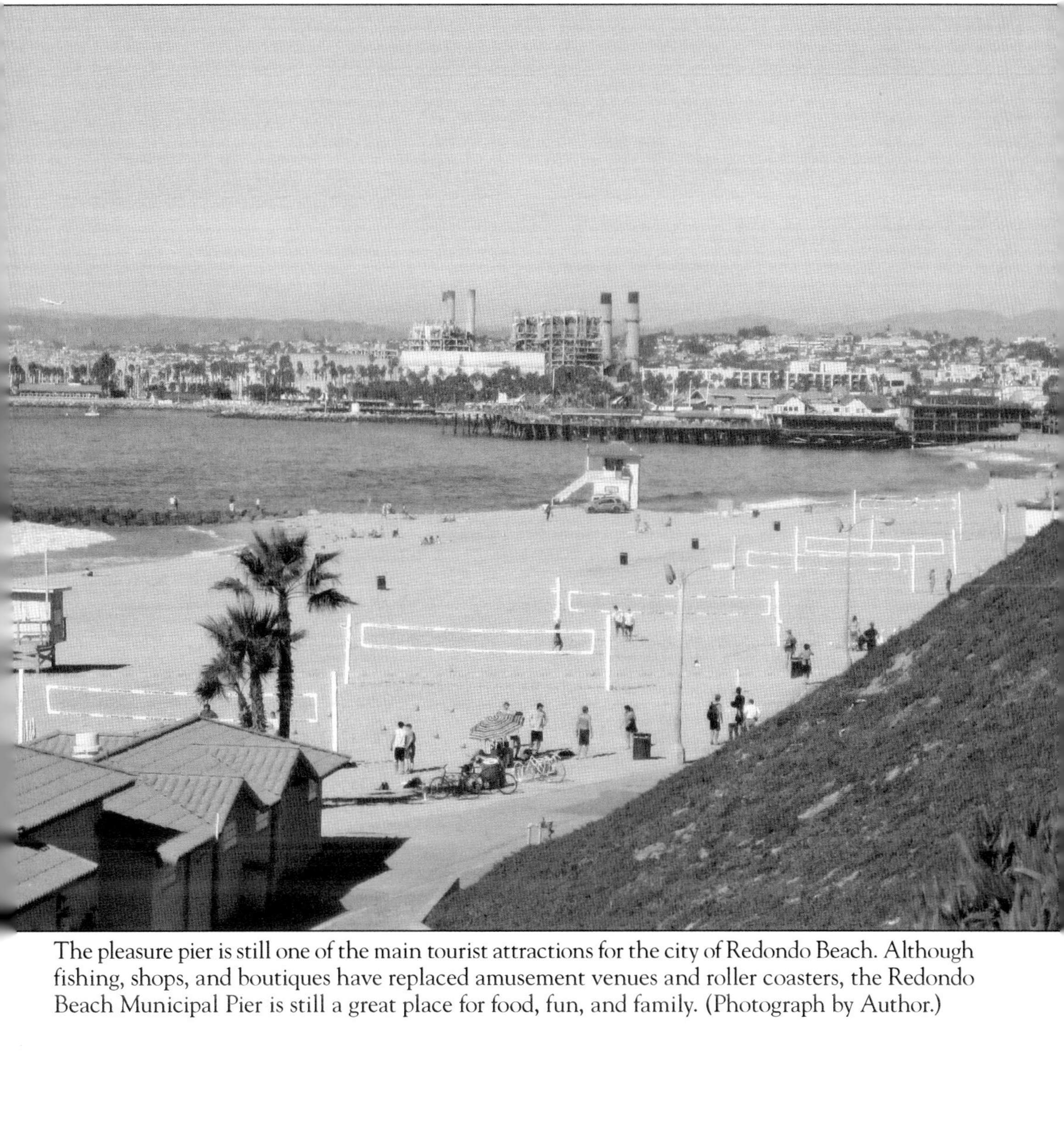

The pleasure pier is still one of the main tourist attractions for the city of Redondo Beach. Although fishing, shops, and boutiques have replaced amusement venues and roller coasters, the Redondo Beach Municipal Pier is still a great place for food, fun, and family. (Photograph by Author.)

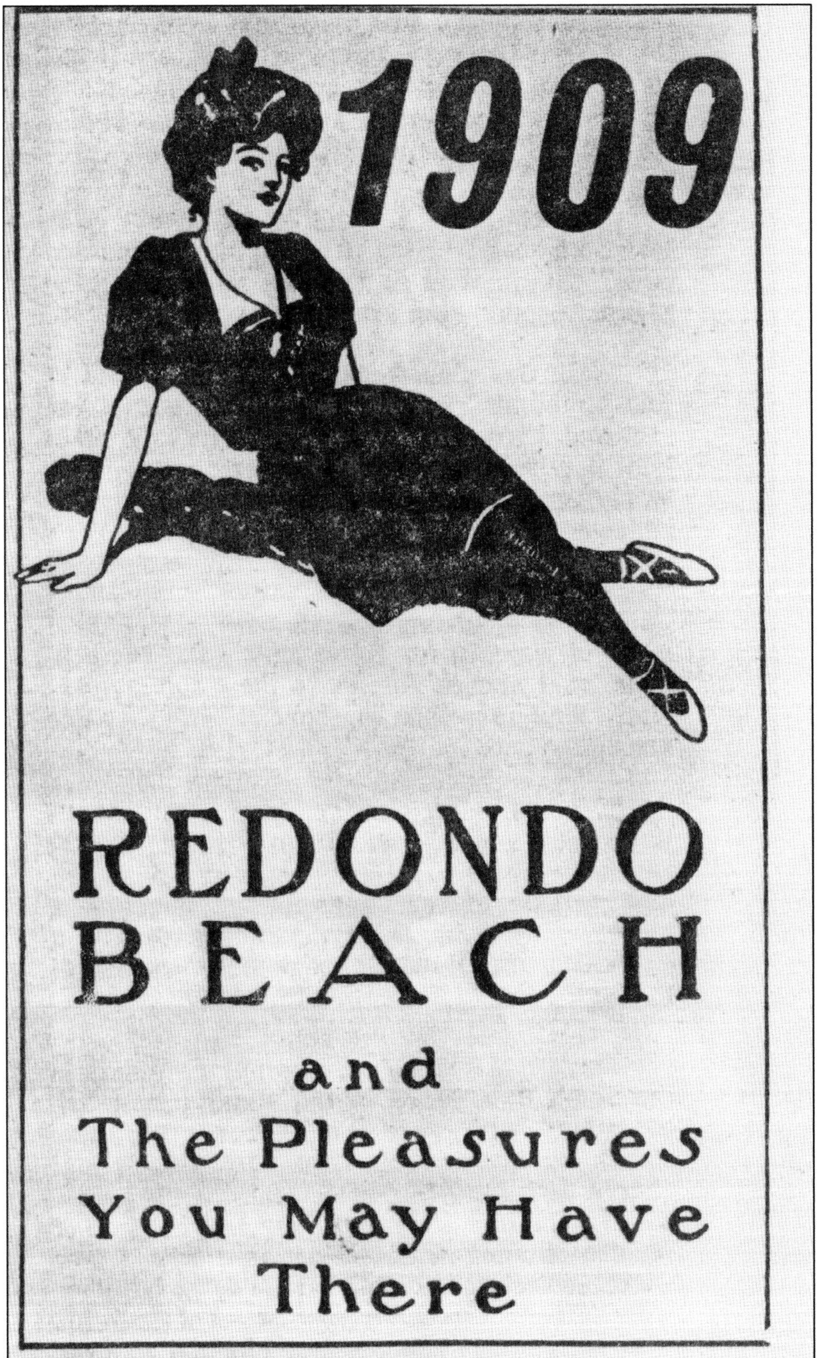

Cover of the first Redondo Beach brochure in 1909 as an appeal to tourists throughout the country.

Above is a 1909 advertisement to visit Redondo Beach. More than 100 years later, Redondo Beach still appeals to tourists and vacationers from all over the world, however, it is easier to get a tan wearing today's swimwear.

Bibliography

"Exterior and Interior Views of Magnificent New Bathhouse and Plunge at Redondo Beach." *Los Angeles Herald*, July 4, 1909, p.12.

Historical Commission of Redondo Beach. Postcard History Series: *Redondo Beach 1880–1930*. San Francisco: Arcadia, 2005.

Johnson, Ken. *Fun, Frustration and Fulfillment: An Historical Study of the City of Redondo Beach*, 1965.

Krintz, Jennifer. *Pleasure Piers & Promenades: The Architecture of Southern California's Early Twentieth Century Beach Resorts*. Athens: University of Georgia, 2009.

McCandless, Michael H. *Well, at Least We Tried: The Seaport of Redondo Beach from 1888 to 1912*. San Diego.

"Redondo Beach Opens Its New Pleasure Pier." *Los Angeles Times*, August 6, 1916, p. I11.

"Redondo Beach Pier Wonder of Engineers." *Los Angeles Times*, October 11, 1915, p. I16.

"Redondo Beach Season Opens with Dedication." *Los Angeles Times*, April 7, 1913, p I18.

Shanahan, Dennis. *Old Redondo: A Pictorial History of Redondo Beach, CA*. Redondo Beach: Legends Press, 1982.

"Twenty Thousand Go To Redondo Beach." *Los Angeles Times*, June 23, 1913, p. I18.

Van Vleet, Miranda. *The American State Fair: Architecture and Preservation*. Athens: University of Georgia, July 2008.

"Will Provide Amusements: Redondo Beach to Have Roller Coaster and Extensive Playground to Divert Its Visitors." *Los Angeles Times*, March 22, 1912, p. I14.